T0211714

Sustainable Urban Agriculture and Food Planning

As urban populations rise rapidly and concerns about food security increase, interest in urban agriculture has been renewed in both developed and developing countries. This book focuses on the sustainable development of urban agriculture and its relationship to food planning in cities.

It brings together the best revised and updated papers from the Sixth Association of European Schools of Planning (AESOP) conference on Sustainable Food Planning. The main emphasis is on the latest research and thinking on spatial planning and design, showing how urban agriculture provides opportunities to develop and enhance the spatial quality of urban environments. Chapters address various topics such as a new theoretical model for understanding urban agriculture, how urban agriculture contributes to restoring our connections to nature, and the limitations of the garden city concept to food security. Case studies are included from several European countries, including Bulgaria, France, Germany, Italy, Netherlands, Romania, Spain, Turkey and the United Kingdom, as well as Australia, Canada, Cameroon, Ethiopia and the United States (New York and Los Angeles).

Rob Roggema is Owner/Director of Cittaideale, Research Office for Adaptive Planning and Design (www.cittaideale.eu) in Wageningen, the Netherlands. He is also Adjunct Professor of Planning with Complexity at the Centre for Design Innovation, Swinburne University of Technology in Melbourne, Australia. He is Director of the Regional Resilience Research Institute, a global institute to help regions becoming more resilient, physically, socially and mentally. Until 2015 he was Professor of Design for Urban Agriculture.

Routledge Studies in Food, Society and Environment

For further details please visit the series page on the Routledge website:
www.routledge.com/books/series/RSFSE/

Sustainable Urban Agriculture and Food Planning

Edited by Rob Roggema

Routledge
Taylor & Francis Group

LONDON AND NEW YORK

earthscan
from Routledge

First published 2016 by Routledge

2 Park Square, Milton Park, Abingdon, Oxfordshire OX14 4RN
52 Vanderbilt Avenue, New York, NY 10017

Routledge is an imprint of the Taylor & Francis Group, an informa business

First issued in paperback 2018

British Library Cataloguing-in-Publication Data
A catalogue record for this book is available from the British Library

Library of Congress Cataloging in Publication Data
Names: Roggema, Rob, editor.
Title: Sustainable urban agriculture and food planning / edited by Rob
Roggema.
Other titles: Routledge studies in food, society and environment.
Description: London ; New York : Routledge, 2016. |
Series: Routledge studies in food, society and environment | Includes
bibliographical references and index.
Identifiers: LCCN 2016004013| ISBN 9781138183087 (hbk) | ISBN
9781315646039 (ebk)
Subjects: LCSH: Urban agriculture. | Sustainable development. | Food
security.
Classification: LCC S494.5.U72 S86 2016 | DDC 630.9173/2—dc23LC
record available at https://lccn.loc.gov/2016004013

ISBN: 978-1-138-18308-7 (hbk)
ISBN: 978-0-367-17281-7 (pbk)

Typeset in Bembo
by FiSH Books Ltd, Enfield

Contents

Illustrations

Figures

Tables

Boxes

Notes on contributors

Kumru Arapgirlioğlu PhD, City and Environmental Planner, Bilkent University, Department of Urban Design and Landscape Architecture.

Deniz Altay Baykan PhD, City and Regional Planner, Bilkent University, Department of Urban Design and Landscape Architecture.

Roseline Njih Egra Batcha Department of Geography, University of Yaounde 1, Cameroon.

Steven Buchanan Researcher, Alfred State College, State University of New York.

Grit Bürgow Research and Development coordinator ROOF WATER-FARM, Dep. Urban Design & Urban Development, Institute of Urban & Regional Planning (ISR), Berlin Institute of Technology – TU Berlin.

Joy Carey Independent Consultant in Sustainable Food Systems Planning, and Director of Bristol Green Capital Partnership, Bristol UK.

Natalie A. Hall Researcher, Queens University Belfast.

Bernard Hubert Senior Scientist at INRA and Professor at the EHESS.

François Jégou Director of Strategic Design Scenarios and professor at ENSAV La Cambre Brussels, Belgium.

Andy Jenkins Researcher, Queens University Belfast.

Greg Keeffe Professor of Sustainable Architecture, Queens University Belfast.

Anna Chiara Leardini architect, co-founder of viÆmilia, chef at EVVIVA restaurant, Riccione, Italy.

Howard Lee Lecturer in sustainable agriculture at Hadlow College, Kent, UK; member of Editorial board of International Journal of Agricultural Sustainability.

Angela Million, née Uttke Head of Coordination ROOF WATER-FARM, Chair-holder for Urban Design & Urban Development at the Institute of Urban & Regional Planning (ISR), Berlin Institute of Technology – TU Berlin.

Tal Alon-Mozes Faculty of Architecture and Town Planning, Technion, Israel Institute of Technology.

Natalia Mylonaki Natalia Mylonaki Design Studio.

Claude Napoléone Scientist on economy at the French National Institute for Agricultural Research (INRA).

Rob Roggema Cittaideale, Research Office for Adaptive Design and Planning, Wageningen, the Netherlands and Adjunct Professor Planning with Complexity, Centre for Design Innovation, Swinburne University of Technology, Hawthorn, Australia.

Esther Sanz Sanz PhD candidate on urban studies at the Advanced School for the Social Sciences (EHESS, France) and on geography at the Autonomous University of Madrid (UAM, Spain).

Laura Sasso PhD candidate in Advanced Research in Urban Systems (ARUS) at the University of Duisburg-Essen.

Stefano Serventi Architect, co-founder of viÆmilia, Architect at Global Arquitectura Paisagista, Lisbon, Portugal.

Anja Steglich Research and Communication Coordinator, Department of Urban Design & Urban Development, Institute of Urban & Regional Planning (ISR), Berlin Institute of Technology – TU Berlin.

Bauke de Vries Department of the Built Environment, Eindhoven University of Technology, the Netherlands.

Tom Veeger Department of the Built Environment, Eindhoven.

Arnold van der Valk Full Professor of Land Use Planning at Wageningen University, the Netherlands and University of Technology, the Netherlands.

Maarten Willems Department of the Built Environment, Eindhoven University of Technology, the Netherlands.

Joshua Zeunert Lecturer at Deakin University, Australia.

1 Introduction

On the brink of why and how: sustainable urban food planning grows up

Rob Roggema

Introduction

In recent years many books on urban agriculture, urban farming, food planning or food systems have been published (De Zeeuw and Drechsel, 2015; Miazzo and Minkjan, 2013; Roggema and Keeffe, 2014; Viljoen and Bohn, 2014). However, this book *Sustainable Urban Agriculture and Food Planning*, marks a shift in perspective. Many discussions and the majority of the research in the past dealt with food safety and security. This book looks beyond these 'why' questions of the food issue, as the questions of 'how' to produce enough, healthy, sustainable and acceptable food close to where it is consumed and where it can be experienced, become more and more important.

Trends

Several trends regarding local and regional production of food can be identified: the scale of urban agriculture, the role of developing countries, the spatial impacts and conditions, the design outcomes, the availability of space, new concepts and new roles for the consumer.

Thinking at the city region scale

Several food-planning scales are currently used to determine the growth of food in or near urbanised areas (see Table 1.1):

1   The city region food system encompasses the complex network of actors, processes and relationships to do with food production, processing, marketing and consumption that exist in a given geographical region that includes a more or less concentrated urban centre and its surrounding peri-urban and rural hinterland – a regional landscape across which flows of people, goods and ecosystem services are managed (FAO and RUAF, 2015).
2   Food system planning is seen as an urban system (Pothukuchi and Kaufman, 1999), however the local scale is not the only scale to look at the food system (Born and Purcell, 2006), as the system is scalable, and can be analysed at higher scales, even global.

3   Urban Agriculture is defined as:

> an industry located within (intra-urban) or on the fringe (peri-urban) of a town, an urban centre, a city or metropolis, which grows or raises, processes and distributes a diversity of food and non-food products, reusing mainly human and material resources, products and services found in and around that urban area, and in turn supplying human and material resources, products and services largely to that urban area.
>
> (Mougeot, 1999)

Urban farming is the growing, processing and distribution of food or livestock within and around urban centres with the goal of generating income (Poulsen and Spiker, 2014; Thoreau, 2010).

Street food is ready-to-eat food or drink sold in a street or other public place, such as a market or fair, by a hawker or vendor, often from a portable food booth, food cart or food truck (Simopoulos and Bhat, 2000).

A street vendor is a person who offers goods or services for sale to the public without having a permanently built structure but with a temporary static structure or mobile stall (or head-load). Street vendors could be stationary and occupy space

Table 1.1 Types of urban food production and their typical scale

Type	Definition	Scale
City region	A more or less concentrated urban centre and its surrounding peri-urban and rural hinterland	Regional landscape
Food system planning	Planning of the food system at the urban or the local scale. The system is scalable	Urban region
Urban agriculture	Agriculture within (intra-urban) or on the fringe (peri-urban) of a town, an urban centre, a city or metropolis	Urban and peri-urban
Urban farming	Farming within and around urban centres	Urban centres
Street food	Food sold in a street or other public place	Street, public/private space
Street vendor	Person selling food on the pavements or other public/private areas, or mobile	Pavement, public area
Community garden	Shared productive land in neighbourhoods, schools, connected to institutions such as hospitals, and on residential housing grounds	Piece of land in neighbourhood

Source: Roggema and Spangenberg (2015)

on the pavements or other public/private areas, or could be mobile and move from place to place carrying their wares on push carts or in cycles or baskets on their heads, or could sell their wares in moving buses (MHUPA, 2004; Sundaram, 2008).

Community garden/consumer collectives: a community garden is any piece of land gardened by a group of people, utilising either individual or shared plots on private or public land. The land may produce fruit, vegetables and/or ornamentals. Community gardens may be found in neighbourhoods, schools, connected to institutions such as hospitals, and on residential housing grounds (University of California, undated).

Despite an increase in local low-scale urban farming projects, such as rooftop gardens, community gardens and mobile street food entrepreneurs, a general trend to start looking at the food system at the city-region scale is visible. For instance the work of FAO/RUAF (Food and Agriculture Organization of the United Nations/Resource Centres on Urban Agriculture and Food Security) (FOA and RUAF, 2015) and IUFN (International Urban Food Network) (Jennings *et al.*, 2015) makes clear that at this level the gains in terms of sustainability, health and efficiency could be large. At this scale the urban metabolism, or the flows of resources inside and outside of the food system, is an important issue and is very promising.

Many of the chapters in this publication emphasise the city region (Chapters 3: Leardini and Serventi; 4: Van der Valk; 5: Sanz Sanz *et al.*; 6: Keeffe *et al.*; and 14: Batcha) or food system (Chapters 2: Keeffe; 3: Leardini and Serventi; 4: Van der Valk; 6: Keeffe *et al.*; and 15: Lee) scales. The urban agriculture (Chapters 2: Keeffe; 7: Zeunert; 8: Mylonaki; 11: Kumru Arapgirlioğlu and Altay Baykan; and 12: Sasso) and urban farming (Chapters 2: Keeffe; 4: Van der Valk; 9: Million *et al.*; and 10: Buchanan) scales are also widely used in this book, while street food and vendors (Chapter 16: Jégou and Carey, regarding intermediate entrepreneurs sale in schools and land markets) and community gardens (Chapter 4: Van der Valk; and 13: Tal Alon Mozes) are only sparsely mentioned (see Table 1.2).

Table 1.2 The scales discussed in the different chapters of this book

Scale	Chapters
City region	3 (Leardini and Serventi), 4 (Van der Valk), 5 (Sanz Sanz *et al.*), 6 (Keeffe *et al.*), 14 (Batcha)
Food system	2 (Keeffe), 3 (Leardini and Serventi), 4 (Van der Valk), 6 (Keeffe *et al.*), 15 (Lee)
Urban agriculture	2 (Keeffe), 7 (Zeunert), 8 (Mylonaki), 11 (Kumru Arapgirlioğlu and Altay Baykan), 12 (Sasso)
Urban farming	2 (Keeffe), 4 (Van der Valk), 9 (Million *et al.*), 10 (Buchanan)
Street food	16 (Jégou and Carey)
Street vendor	16 (Jégou and Carey)
Community garden	4 (Van der Valk), 13 (Tal Alon Mozes)

The role of developing countries

In Dar es Salaam and Nairobi, just to name a couple of cities in developing countries, the growth of food in urban areas and slum areas is a common phenomenon (Conway, undated; Foeken and Mwangi, undated; Foeken *et al.*, 2004; Jacobi *et al.*, undated; Kenyan Ecotourist, 2012; Lee-Smith, 2013; Mayoyo, 2015; Schmidt, 2011). Increasingly it becomes clear that these cities should not only be seen as places where urban farming methodologies and techniques developed in developed countries could be implemented, but these cities have a large experience in organising, implementing and growing food close to the consumers. Besides the still-necessary support for the poorest people in arranging their local food supply, including set up of urban agriculture projects, these cities should also be approached as a knowledge base to learn from. The experiences in Dar es Salaam and Nairobi are widespread, as several chapters in this publication illustrate. Especially in Chapter 13, Batcha discussed the situation in Cameroon.

Spatial impacts and conditions

In urban agriculture specific fields of research have been distinct. There is a huge body of knowledge about the resource efficiency and environmental performance of urban agriculture projects (Allen, 2003; Deelstra and Girardet, 2000; Mougeot, 2010), and at the same time many scholars have studied the social impacts of these projects (De Bon *et al.*, 2010; Mougeot, 2010; Nugent, 2000) or their sustainability (Koc, 1999; Pearson *et al.*, 2010; Smit *et al.*, 1996). So far, these topics have mainly been looked at from a sectorial perspective. In the current timeframe there is an increase in studies and projects that observe urban agriculture from one integrated frame. The studies carried out in Rotterdam for instance show the integration of spatial needs of urban food production with the spatial conditions and potentials in the city (De Graaf, 2011). Four types of urban agriculture (forest gardening, small plot intensive farming (SPIN), roof hydroponics and aquaponics) are matched with the potentials and constraints (soil, water, heat islands and organic waste) in the city, which leads to an integrated vision on the chances for urban agriculture in Rotterdam (see Figure 1.1) though the number of factors and types is limited.

What can be distinguished is that the approach to urban agriculture is increasingly integrating different topics into one frame. In such a frame (see Figure 1.2) design aspects (scales, design principles, concepts and strategies, potentials, existing spatial structures and patterns), environmental parameters (urban metabolism, flows of water, nutrients and energy) and economic (business models), social (inclusion, cohesion) and agricultural (productivity) factors are factored in the framework (Roggema, 2014).

The framework illustrated in Figure 1.2 consists of two halves. To the left hand side the agricultural productivity aspects are located. The productivity depends on the demand (size of population, diet), the feasibility of crop types and the economic system. These three factors determine the agricultural system. To the right of the

Figure 1.1 Opportunity map: room for urban agriculture in Rotterdam

Source: Paul de Graaf Research & Design (2011).

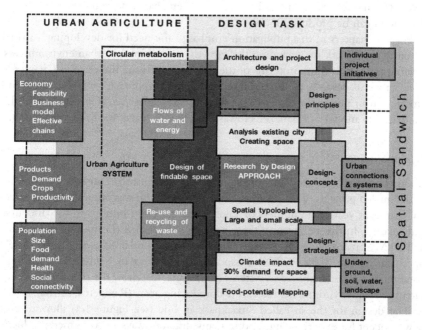

Figure 1.2 Integrated framework for urban agriculture

Source: Roggema (2014).

framework the design tasks are located. These tasks are divided in three levels of scale. Some of the tasks are effective at a strategic level. Here we are talking about the food potentials of a certain area, which may be under the influence of climatic impacts. The underground, soil, water system and the landscape determine the potential to grow food. Design strategies are effective at the city-region level. The design concepts are effective at the urban level. Spatial typologies and existing spaces determine the type of urban agriculture projects that can be implemented at this level. Urban patterns and structure determine the design. The lowest level is the design principle, which is effective at the project level. For park and garden designs these spatial principles are used to create a concrete design.

The two halves are connected with each other through the flows of energy, resources and water, which work both in the agricultural system as well as in the designs at several levels. The design and planning of the circular metabolism allows us to find spaces and locations where best to grow food in urban environments. The availability of flows of water, energy and nutrients for instance is necessary to grow food in places that are identified through design at different scales. At the same time, the availability of these resources is essential to meet the economic and food demands of the regional population. The regional agricultural system cannot function without sufficient resources. Therefore, the design and identification of spaces for food are only effective when supplied with these resources and this makes it possible to supply food for the local demands. Working in this framework helps to create vision and design in an integrated and holistic way. This implies that when the framework is used there is a greater chance that more food that is demanded can be produced locally.

Several chapters in this publication emphasise the need for developing a spatial systemic framework for including the growth of food in our urban environments. Keeffe proposes to think in terms of the hardware software interface in Chapter 2, Sanz Sanz *et al.* describe a GIS-based methodological approach for typecasting food production in peri-urban areas (Chapter 5) and Keeffe *et al.* (Chapter 6) use the Garden City model of Ebenezer Howard to identify spaces for food production.

Design outcomes

Another visible trend in discussing food production in urban areas is the increase of design-led projects. The importance of a good design was often underestimated, but in recent years the numbers of valuable design contributions to the discourse is increasing. In this publication the Chapters 2 (Keeffe), 3 (Leardini and Serventi) and 8 (Mylonaki) illustrate this development. The four designs that were developed during the Sixth AESOP (Association of European Schools of Planning) conference on sustainable food systems (see Figure 1.3) show the transformation of intense urban environments into food-0producing areas (Roggema, 2015a).

The design for the so-called Smaakpark in Ede (see Figure 1.4) illuminates a new concept for experiencing food in many different ways: as consumer, producer, holidaymaker, playground, wedding place and cooking studio and restaurant.

And there are many others, such as the Zuidpark in Amsterdam, where food

Figure 1.3 Four designs for intensive productive urban landscapes in Groningen, Veghel, Amsterdam and Leeuwarden

Source: Roggema (2015a).

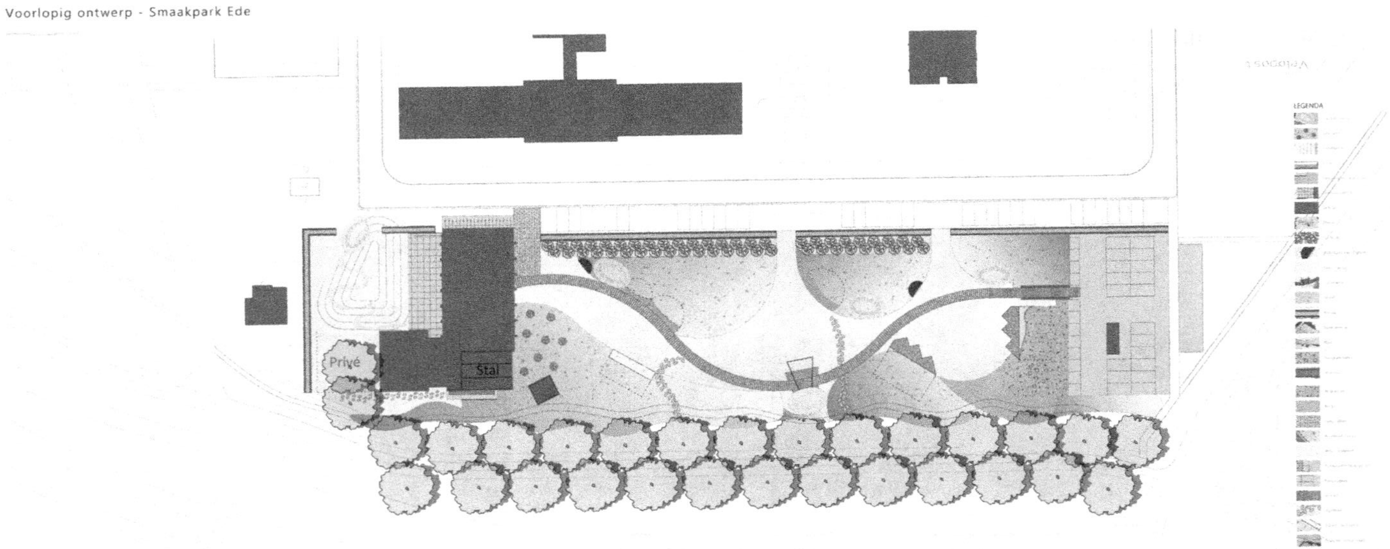

Figure 1.4 Design for Smaakpark Ede

Source: Weij *et al.* (2016).

grows in an office environment, the food forest in the surroundings of Vlaardingen and the design for the Floriade area in Almere (see Figure 1.5).

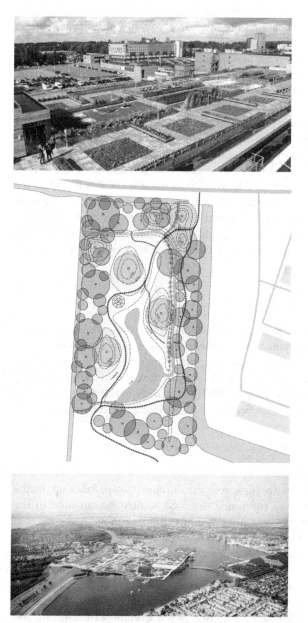

Figure 1.5 The Zuidpark in Amsterdam, Food Forest Vlaardingen and the Floriade in Almere

Source: Food Forest Vlaardingen from Paul de Graaf Research & Design/Rotterdam Forest Garden Network (2015); Floriade in Almere ©MVRDV.

Is there enough space available?

There is an increasing debate about the question whether there is enough space for growing food in the city. Many initiatives in urban areas have been realised, but when we calculate the real impact on the food supply within urban boundaries a tiny 0.002 per cent of food consumption can be produced (Roggema, 2014, 2015b). This can be seen as a weak performance. Of course there are many other reasons to grow food in the city. For instance environmental benefits or social connectivity are important factors. But when the contribution to food supply is close to zero, the question of whether there is enough space available is valid. Not much research is focusing on this subject yet. For the Amsterdam area, recent research found that 12.5 per cent of the surface area (without counting roofs, valuable ecological spaces or private areas) of the city is easily transformable into food productive space. This could provide 25 per cent of the population of Amsterdam with vegetables, herbs and fruits (Mulder and Oude Aarninkhof, 2014). When we take this result and include the potential of roofs, underground spaces, private areas and space inside buildings, the number could be raised to 90 per cent (Roggema, 2015b). However, this implies a diet of vegetables and fruits. The different diets and their spatial consequences is nicely illustrated by comparing Howard's Garden City model in Chapter 6 (Keeffe *et al.*).

New food experiences

In addition to the growth of food and its distribution to the consumers, food is increasingly seen as an experience. This includes the design of food, food safaris, pop-up restaurants, food festivals such as the capital of taste (see www.puur-e.nl) or neighbourhood food nights, such as in east Amsterdam (see www.foodnight.nl) and many others. In this publication, Chapter 10 (Buchanan) gives insights in the possibilities for new experiences in the fields of education and recreation in Queens, New York.

The role of the consumer

A last trend to be signalled is the new role the consumer often takes up. In the recent past the consumer only consumed the food, which was on offer in the supermarket. There was hardly any other option than to buy food and cook it at home or eat food in a restaurant. Now consumers maintain public food gardens, harvest the produce and cook it at home. Examples of this are the *pluktuinen* (pick-gardens), such as Pluktuin de Bosrand in Wageningen (www.pluktuindebosrand.nl) or Pluktuin De Kruudhof in Halle (http://dekruudhof.nl). Also, consumers can eat fresh food from locally produced and cooked food, such as offered by street food vendors (Boer Geert, Vleesch noch Visch (www.vleeschnochvisch.com) and Tho Vietnamese Loempia's). The third new role for consumers is when they take up the role of (professional) producer. The consumer is also an entrepreneur and capable of delivering agreed amounts and qualities of produce to restaurants, street vendors or (super)markets. In Chapter 15, François Jégou and Joy Carey explore the new intermediate role in between consumers and producers.

Proposition: a new definition

As becomes clear in the trends described above, food planning is entering a new stage of its development. Instead of conducting the discussion about the necessity to provide safe and secure food, this stage is putting at its centre how to enjoy and produce healthy and environmentally friendly food. The attention on the larger urban and regional scales and the design approaches emphasises the tactile and conceptual spatial side of urban food planning, including discussion about the availability of sufficient spaces in the city to grow food. The knowledge available in developing countries, the changing role of consumers and the trend of food experiences all imply social interactions in actively cultivating crops in sometimes confined places in urban regions. The latter trends are also focusing the debate on how to grow and enjoy food production, rather than problematising the issue. Therefore, the chapters in this publication altogether give reason to adjust the definition of food planning to become:

> *Food planning provides the spatial conditions to produce and experience enough, sustainable, healthy and acceptable food*

This new definition consists of the following ingredients:

Spatial conditions: Food planning should create, design and safeguard the spatial possibilities to grow food in amounts that roughly meet the demands of the urban population, at least for the vegetable, herb and fruit components of the food pallet.

Produce and experience: Food planning should provide the places where food can be produced, but also where it can be experienced. Therefore food spaces must be productive (a wide range of fruits, vegetables and herbs can be grown) and at the same time accessible for co-producers, tourists and consumers to enjoy the production, maintenance, harvesting, cooking and consumption of local food.

Enough: Food planning should arrange available spaces to produce the produce that the local urban population demands.

Sustainable: Food planning should create the spatial coherence in food production that makes it possible to reuse and recycle resources (energy, water) and materials (nutrients, waste) in the production process, and arranges a connected system with short transport links.

Healthy: Food planning provides the possibility to produce food without using pesticides and other artificial products, and offers fresh food at close distances.

Acceptable: The food planning system provides food that is sustainable, but also culturally responsible. In times of migration the cultural mixes in the population are increasing the range of food diets, menus and crops produced. Food planning needs to offer the opportunities for each of the cultural demands to deliver the specific produces. Food, cooking and eating is an important basis for sharing cultural differences and joining a common world.

Conclusion

In this book a range of chapters are written about food planning. Coming from a range of directions, the common message of these chapters is a positive one: if we change the way we produce food we are capable of providing food for everyone in a sustainable and enjoyable way. Each of the chapters contributes to thinking of solutions for the current food issues, without becoming too theoretical. The stories incorporate a hands-on attitude to discovering how to grow food in a sustainable way, close to where it is enjoyed.

The trends mentioned above mark a shift towards constructive thinking in food planning. This shift is just starting and requires further research, practical projects and continuous knowledge exchange between consumers, researchers, producers, practitioners and policymakers. This is necessary because the current amounts of food grown within or close to urban environments are still not sufficient to feed a reasonable part of the population. The available space in the city, yet undiscovered, needs to become visible and take up a role in the food system. As long as we only trust in large-scale, efficient yet unsustainable productive agriculture, the major food issues will not be solved in the long term. It is necessary to develop city-regional food systems in which a large number of beautiful productive spaces are designed that are capable of growing food for the majority of the population. If this book can contribute to realising these city-region food systems all over the world it has reached its goal.

References

Allen, A. (2003) Environmental planning and management of the peri-urban interface: perspectives on an emerging field. *Environment and urbanization* 15(1), 135–148.

Born, B. & Purcell, M. (2006) Avoiding the local trap. Scale and food systems in planning. *Research Journal of Planning Education and Research* 26, 195–207.

Conway, K. (undated) *CASE STUDY: Dar es Salaam, Tanzania — Building the food-secure city: Incremental progress brings about change*. IDRC. Online: www.idrc.ca/EN/Resources/Publications/Pages/ArticleDetails.aspx?PublicationID=541.

Deelstra, T., & Girardet, H. (2000). Urban agriculture and sustainable cities. In: Bakker, N., Dubbeling, M., Gündel, S., Sabel-Koshella, U., & de Zeeuw, H. (eds) *Growing Cities, Growing Food. Urban agriculture on the policy agenda*. Feldafing, Germany: Zentralstelle für Ernährung und Landwirtschaft (ZEL), pp. 43–66.

De Bon, H., Parrot, L., & Moustier, P. (2010). Sustainable urban agriculture in developing countries. A review. *Agronomy for Sustainable Development* 30(1), 21–32.

De Graaf, P. (2011) *Ruimte voor Stadslandbouw in Rotterdam*. Rotterdam: Stimuleringsfonds voor Architectuur.

De Zeeuw, H., & P. Drechsel (eds) (2015) *Cities and Agriculture: Developing Resilient Urban Food Systems*. London: Routledge.

FAO & RUAF (2015) *City Region Food Systems. Building sustainable and resilient city regions.* Rome: FAO, RUAF.

Foeken, D., Sofer, M., & Mlozi, M. (2004) *Urban Agriculture in Tanzania: Issues of Sustainability.* African Studies Centre Research Report 75 / 2004. Leiden: Leiden University.

Foeken, D., & Mwangi, A.M. (undated) Increasing food security through urban farming in Nairobi. Online: https://openaccess.leidenuniv.nl/bitstream/handle/1887/4670/ASC-1241504-038.pdf?sequence=1.

Jacobi, P., Amend, J., & Kiango, S. (undated) Urban agriculture in Dar es Salaam: providing an indispensable part of the diet. RUAF. Online: www.ruaf.org/sites/default/files/DaresSalaam_1_1.PDF.

Jennings, S., Cottee, J., Curtis, T., & Miller, S. (2015) *Food in an Urbanised World: The role of city region food systems in resilience and sustainable development.* Paris: IUFN, The Prince's Charities and 3keel.

Kenyan Ecotourist (2012) *Breaking off the poverty chains: Urban farming in Nairobi, Kenya.* Online: https://gcardblog.wordpress.com/2012/10/20/breaking-off-poverty-chains-case-urban-farming-nairobi-kenya/

Koc, M. (1999). *For Hunger-proof Cities: Sustainable urban food systems.* Ottawa: IDRC.

Lee-Smith, D. (2013) *Changing Lives – Urban Farmers of Nairobi, Kenya.* Online: www.cityfarmer.info/2013/12/30/changing-lives-urban-farmers-of-nairobi-kenya/.

Mayoyo, P. (2015) *How to grow food in a slum: lessons from the sack farmers of Kibera.* Online: www.theguardian.com/global-development-professionals-network/2015/may/18/how-to-grow-food-in-a-slum-sack-farmers-kibera-urban-farming.

MHUPA (2004) *National policy on Urban Street Vendors. Department of Urban Employment and Poverty Alleviation.* Ministry of Urban Development and Poverty Alleviation. New Delhi: Government of India.

Miazzo, F., & M. Minkjan (eds) (2013) *Farming the city: Food as a tool for today's urbanisation.* Haarlem: trancityxvaliz.

Mougeot, L.J.A. (1999) *Urban Agriculture: Definition, Presence, Potential and Risks, Main Policy Challenges.* CFP report Series, Canada.

Mougeot, L.J.A. (ed.) (2010) *Agropolis: The Social, Political and Environmental Dimensions of Urban Agriculture.* London: Routledge.

Mulder, M. & Oude Aarninkhof, C. (2014) *Stadslandbouwdoos.* Den Haag: Atelier Rijksbouwmeester.

Nugent, R. (2000). The impact of urban agriculture on the household and local economies. In: Bakker, N., Dubbeling, M., Gündel, S., Sabel-Koshella, U., & de Zeeuw, H. (eds) *Growing Cities, Growing Food. Urban agriculture on the policy agenda.* Feldafing, Germany: Zentralstelle für Ernährung und Landwirtschaft (ZEL), pp. 67–95.

Pearson, L.J., Pearson, L., & Pearson, C.J. (2010). Sustainable urban agriculture: stocktake and opportunities. *International Journal of Agricultural Sustainability* 8(1–2): 7–19.

Pothukuchi, K., & Kaufman, J.L. (1999) Placing the food system on the urban agenda: The role of municipal institutions in food systems planning. *Agriculture and Human Values* 16: 213–224.

Poulsen, M.N., & Spiker, M.L. (2014) *Integrating Urban Farms into the Social Landscape of Cities. Recommendations for Strengthening the Relationship between Urban Farms and Local Communities.* Johns Hopkins Bloomberg School of Public Health.

Roggema, R. (2014) Finding spaces for productive cities. In: Roggema, R., & G. Keeffe (eds) (2014) *Why We Need Small Cows. Ways to Design for Urban Agriculture.* Velp: VHL Press, pp. 37–61.

Roggema, R. (2015a) The reinvention of the academic conference: how active delegates develop productive city concepts. *Future of Food: Journal on Food, Agriculture and Society* 3(1): 63–78.

Roggema, R. (2015b) Towards fundamental new urban planning for productive cities: the quest for space. *Proceedings Agriculture in an Urbanising Society Conference*. Rome, 17–19 September.

Roggema, R., & G. Keeffe (eds) (2014) *Why We Need Small Cows. Ways to Design for Urban Agriculture*. Velp: VHL Press.

Roggema, R., & J. Spangenberg (2015) New urban networks for linking the urban food production-preparation-consumption chain. *Proceedings 51st ISOCARP conference*. Rotterdam/Wageningen, 19–23 October.

Schmidt, S. (2011) Case study #7-12: Urban agriculture in Dar es Salaam, Tanzania. In: Per Pinstrup-Andersen & Fuzhi Cheng (eds) *Food Policy for Developing Countries: Case Studies*. Online: http://cip.cornell.edu/dns.gfs/1297701745.

Simopoulos, A.P., & Bhat, R.V. (2000) *Street Foods*. Basel: Karger AG.

Smit, J., Nasr, J., & Ratta, A. (1996). *Urban Agriculture: Food, jobs and sustainable cities*. New York: The Urban Agriculture Network.

Sundaram, S.S. (2008) National policy for urban street vendors and its impact. *Economic and Political Weekly 43*: 22–25.

Thoreau, C.M. (2010) *Defining Urban Farming*. Vancouver Urban Micro. Online. URL: http://urbanmicro.ca/2010/10/15/defining-urban-farming/.

University of California (undated) *Community gardens. What is a community garden?* Online: http://ucanr.edu/sites/MarinMG/Community_Service_Projects/Marin_Community_Gardens/.

Viljoen, A., & K. Bohn (2014) *Second Nature Urban Agriculture. Designing Productive Cities*. London: Routledge.

Weij, C., Roggema, R., & Vermeend, T. (2016) *Ontwerp voor het Smaakpark, presentatie van het voorlopig ontwerp*. 5 April. Ede: Puur-e.

2 Hardware software interface

A strategy for the design of urban agriculture

Greg Keeffe

Introduction

This chapter describes how urban agriculture differs from conventional agriculture not only in the way it engages with the technologies of growing, but also in the choice of crop and the way these are brought to market. The author proposes a new model for understanding these new relationships, which is analogous to a systems view of information technology, namely the hardware software interface (HSI) approach.

The first component of the system is hardware. This is the technological component of the agricultural system. Technology is often thought of as equipment, but its linguistic roots are in 'technis', which means 'know how'. Urban agriculture has to engage new technologies, ones that deal with the scale of operation and its context, one, which is different than rural agriculture. Often the scale is very small and soils are polluted; here the idea of 'technology' in urban agriculture could be technical such as aquaponic systems, or could be soil-based agriculture such as allotments, window boxes or perhaps methods based on permaculture. The choice of method does not necessarily determine the crop produced or its efficiency. The crop produced is linked to the biotic components that are added to the hardware; this is seen as the 'software' in the system.

The software in this model of urban farming is the ecological elements within the system. These produce the crop, which may or may not be determined by the technology used. For example, a hydroponic system could produce a range of crops, or even fish or edible flowers. Software choice can be driven by ideological preferences such as permaculture, where companion planting is used to reduce disease and pests, or by economic factors such as the local market at a particular time of the year. The market determines the monetary value of the 'software'. Obviously small, locally produced crops are unlikely to compete against intensive products produced globally, however the value locally might be measured in different ways, and might be sold on a different market. This leads to the final part of the analogy: interface.

The interface is the link between the system and the consumer. In traditional agriculture, there is a tenuous link between the producer of asparagus in Peru and the consumer in Europe. In fact very little of the money spent by the consumer ever reaches the grower. Most of the money is spent on refrigeration, transport and profit for agents and supermarket chains. Local or hyper-local agriculture needs to bypass or circumvent these systems, and be connected more directly to the consumer. This is the interface. In hyper-localized systems effectiveness is often more important than efficiency, and direct links between producer and consumer create new economies.

Background

The development of organized agriculture and the rise of the city form are directly connected from as early as 6000 BC (or earlier) in what James Henry Breasted called 'the Fertile Crescent' (Abt, 2011) in places such as Erbil. However particularly since industrialization there has been an increasing separation between the places of production and those of consumption. This has come about through a radical re-imagining of the farm as a hyper-industrialised system at an un-bodied scale. Today a city such as London needs around 150 times its own footprint just to feed itself (Best Foot Forward *et al.*, 2005). This disconnection in scale is compounded by the fact that this productive land is no longer nearby, but is scattered piecemeal around the globe: wheat may come from the USA; beef from South America; fruit from Africa and Asia; fish from Australasia and so on. This scattering is a direct result of international trade agreements that have liberalized flows based on supply cost, rather than on ecological factors.

However it is not just food that is distributed far from the city: energy production and the fuel itself are similar in their disconnection. A city such as Liverpool UK, with 1.5 million people in its region, has only a tiny amount of power produced within the city. This consists of a small but growing fraction of off-shore wind, but there is little else produced, even the nearest fossil fuelled electricity is produced some 20 kilometres from the city, and there is little or no solar electricity generation in city, limited to a small number of domestic installations.

In terms of resilience, this creates a vulnerability to crisis in the contemporary city: through the utilization of the very global system of trade that it created, the city has become more and more dependent on these trade networks for its metabolism. If these networks were to collapse or malfunction, due to fuel shortage, resource depletion or climate change, the situation the city finds itself in could be untenable. Such might be considered rare, but even minor events such as the 'ash cloud' from the Eyjafjallajokull volcano in Iceland during 2010 had critical effects on the city. The cloud prevented flights in and out of the UK for only a few days, but a mere three days after the stoppage, empty shelves were seen in supermarkets such as Tesco, where their just-in-time strategy of food supply was put under pressure. As the head of the UK government's countryside agency, Lord Cameron once said: 'The UK is only nine meals from Anarchy' (Boycott, 2008).

The supply of fuel oil is equally vulnerable; not only can the price be affected by market supply and demand fluctuations, but political events can also halt supplies coming from far afield. The lack of stores of fuel in a process that relies on a 'just-in-time' strategy means that even short events can have a catastrophic effect. A short strike in 2000 by refinery workers and fuel delivery drivers in the UK created petrol shortages after only two days (BBC-News, 2000). The more serious effect of fuel shortage can be seen in Bulgaria, where disagreements with Russia during February 2008 resulted in disruption to the gas supply. This compounded by harsh winter weather left 60,000 dead and raised the concept of 'fuel security' for the first time. Indeed recent issues in the Ukraine have bought this clearly into focus (BBC-News, 2014).

The concepts of food security and fuel security are widely acknowledged and considered, but these are not issues that are easy to remedy, especially when the policies of the last 50 years have been about developing a globalized economy. Arjan Appadurai in his book *Modernity at Large* describes globalization as a system based on flows of global commodities (Appadurai, 1996) and these commodities flow in large market-based networks, which rely on a just-in-time delivery system, with as little (or no) storage as possible. On the face of it, this begs the question, how do we equip our cities and country to be more resilient in the face of this century's problems: climate change, resource depletion, peak oil, increased population and international strife? The author champions a move to a less globally sourced system (at least in part) and this chapter looks in particular to develop strategies for a more localized production of energy and food. The ultimate aim is to develop strategies that not only help to mitigate the effects of the above, but also help to make the city a more livable and equable place.

The chapter describes in detail two projects at very different scales, and although different in scale and output, both projects highlight the need for a process and place-based approach to urban agricultural design. The two projects are both based in post-industrial cities in the UK: the first a proposal for a large scale city-wide installation of an energy landscape growing algae in Liverpool, UK; and the second the design, build and operation of a small, hyper-localized technical food system in Salford, Manchester, UK.

Hardware software interface

The author has argued for some time that urban agriculture is very different in scale and engagement from modern intensive agriculture (Gundlach, 2015). For example, urban agriculture is often concentrated around participatory projects and organic methods, and many of the projects carry a polemic, rather than efficiency, focus. In addition, projects are often consumer-led where local people consume the output locally. Often the scale of implementation is relatively small compared to conventional production methods (even in the Bio-port Project when compared to oil drilling – see below); this means that it is difficult to compete against industrialized production on a purely economic basis. On a financial basis in an open global market, urban agriculture does not make sense; not only is there a large reduction in the economy of scale, but also many of the techniques are a lot less intensive. However there are many other benefits, and it is necessary to consider the implementation in an urban setting in a different way.

The HSI strategy considers urban agriculture not as agriculture in the city per se, but as a multilayered urban design strategy. This urban design strategy sees the integration of agriculture as a holistic urban design problem that is interdisciplinary, and based on networks and agents rather than purely the technical issues of agriculture at a small scale. In this way urban agriculture becomes an important design and engagement tool, which creates new contexts for livability, rather than it being an add-on to a finished master plan.

The strategy is systems based, in that it sees urban food production as a nested system within the city, one that engages with place, people and technology and it looks to integrate the three components of urban agriculture in a way that supports a synergy with its location. This can be seen as a contextual place-based solution that links with ideas of livability, urban metabolism and circularity, resilience and local distinctiveness. Each component of the system has a separate knowledge base, but it is the design and integration of the three elements that creates the effective urban condition as a technical/ecological/social hybrid.

Hardware

To design cities to include urban agriculture needs a way of seeing the practice as a designable and configurable element of the urban fabric, rather like the way designers see a waste water system or urban mobilities. Once considered in this way, urban agriculture becomes urban technology to be applied, rather than purely a method of cultivation that can fit in void space in the city. This liberates the

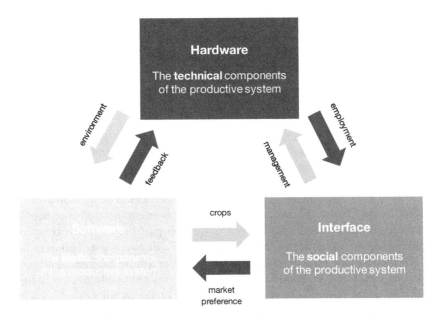

Figure 2.1 Hardware software interface

designer to develop rational and comprehensive ways of developing its incorporation, instead of viewing it as an amateur and non-essential element for which space may need to be provided or might happen by chance.

This holistic reading of urban agriculture breaks down the practice to three separate elements that cover the whole of the socio-technical system. The system can now be understood from different viewpoints: as a part of the urban metabolism; as an eco-system service; or as a consumer choice; or lifestyle, and therefore can be designed. As a hyper-localized socio-technical system, urban agriculture can be considered as three separate components: the growing technology; the choice of crop; and the engagement with the social network of consumption.

The first component of this system is called 'hardware', the growing technology to be used. This is the technological component of the agricultural system. Technology is often thought of as equipment, but its linguistic roots are in 'technis', which means 'know how'. Thus technology in agriculture could be technical such as raceway ponds, as used in Bio-port, or other futuristic systems such as aquaponics, aeroponics or hydroponics. Equally however the technology could be soil based, say an agriculture utilizing a compost-based system such as used in an allotment, window box or permaculture system, or perhaps a crop-rotation system. The choice of method does not necessarily determine the crop produced or its efficiency. This is linked to the biotic that is added to the hardware, which I call the 'software'. Choice of technology will depend on a reading of place and an assessment of appropriateness. Indeed, one technology is not superior or higher than another; aquaponic systems have been used successfully in *favelas* (Roggema, 2014; Roggema *et al.*, 2014), just as compost-based systems are still being applied in comparatively wealthy English towns such as Todmorden, Yorkshire (Paull, 2013).

Software

The biotic element of the system is seen as the 'software'. This component allows the system to be adapted over time to cater for changes in season, climate or market demand. The biotic output of the system may or may not be determined by the hardware. For example, an aquaponic nutrient film system could produce a range of leaf crops or perhaps produce edible flowers, which may vary through the seasons or because of shifts in local demand or market price. The software allows the fixed technological system to adapt to the context in which it operates: agriculture has always been adaptable in this way, but the methodology allows the technology to be designed without necessarily knowing to exact output of the system. This allows designers to make strategic decisions, which can be adapted at a later date (or over time).

Indeed, software choice can also be driven by ideological preferences such as permaculture, for example, where companion planting is used to reduce disease and pests, or perhaps by economic factors such as the local market at a particular time of the year. The market, over which the designer has no control, determines this flexibility in the system, allowing adaptation as the monetary value of the 'software' changes. Obviously small, locally produced crops are unlikely to compete against industrialized products, in a global market, but their direct value locally may be very different, also the small scale of the system allows for more rapid adaptation to local forces. This leads to the final part of the analogy: interface.

Interface

The interface is the link between the harvested crop and the consumer, and perhaps beyond if we are to close the cycle. In traditional agriculture, there is a very tenuous link between the producer of, say, apples in Chile and the consumer in Europe: the conditions of employment are unknown and the effect on the environment is similarly unknown. In fact, there is little communication between producer and consumer, even financially. Indeed, very little of the money spent by the consumer ever reaches the grower. Most of the money goes to profit for agents and supermarket chains, and the rest on refrigeration and transport. The refrigeration and transport have another effect: they greatly increase the ecological footprint of the crop through the use of fossil fuels and the delay from harvest to market reduces greatly the nutritional value of the produce. Apples for example can often be stored for up to 12 months before selling (Rickman *et al.*, 2007). Local or hyper-local agriculture can and needs to bypass or circumvent these systems, and be connected more directly to the consumer. This is the interface. In hyper-localized systems, like the Biospheric Project, leaf crops are sold direct to restaurateurs who pay equivalent prices but receive local organic foods. Parts of the crop that do not meet aesthetic standards or are in oversupply are sold through the 78 Steps Wholefood shop direct to the public in the neighbourhood. This creative use of the interface not only maximizes economic return but also makes the project effective socially.

Bio-port, Liverpool, UK: large-scale implementation

This case study looks at the development of an agricultural system that produces energy crops, and shows that the implementation of relatively simple technology at a large scale could transform the productive capacity of a city and aid urban resilience in the long term.

Liverpool

Liverpool was at the start of the twentieth century the world's largest port. Sited on the west coast of the UK, sheltered in the Mersey Estuary, its deep-water port was perfectly positioned between the industrial powerhouse of Manchester and the Americas and India. However since then as trade within Europe has become more widespread and the size of shipping has increased, Liverpool has fallen on hard times. UK government figures show that Liverpool's population has halved since the Second World War, and those that remain are poor and under-educated (Oswalt, 2005).

Throughout the city, dereliction abounds: the once prosperous dockyards covering some 12 km of river frontage are now vacant, and reports show that over 12 per cent of all land in the city is derelict (Save Liverpool Docks, 2008). Even with Category One European Union funding (reserved for the poorest cities and regions), the future is difficult. Geography is working against the city: it is relatively isolated from the prosperous south east of the UK and indeed Europe, and it has no natural resources to fall back on. Even the sea, once its raison d'être is turning against it, as sea level rises and storm surges in the Irish Sea put its World Heritage-status mercantile city centre under threat of deluge.

Recent times have seen decay, poverty and shrinkage at frightening levels. Nearly half of Liverpool's 28 districts are rated as in the worst 50 in the whole of the UK with respect to deprivation, with four in the bottom ten. In some areas adult unemployment is at 90 per cent. In 2004, the Commission for Architecture and the Built Environment (CABE) in the UK reported that 'space is a luxury in the modern city': this cannot be held as true in Liverpool, where shrinkage has created a toxic oversupply of land, which makes space difficult to navigate, and density or lack of it destroys once thriving neighborhoods. Here space could be seen as a liability. Compounding this spatial collapse is societal collapse: jobs are scarce and fuel poverty is a real concern. Any new solution to Liverpool's plight must not only readdress Liverpool's function but also reconfigure the prodigious amount of spare space to remake a contiguous urbanism. The project described here is a radical but realistic attempt to reverse Liverpool's fortunes by an application of new technology perfectly suited to its estuarine position.

Fossil fuel futures

Peak oil has been reached (Roberts, 2010) and yet there seems to be no reduction in demand for fossil fuels. Many of the renewable replacements have issues regarding the necessary continuous supply. Wind energy is only produced when its windy and solar energy only collected during the day. Energy storage on a vast scale is complex and difficult. In the book *After the Car*, Kingsley Dennis and John Urry explain that due to technological 'lock-in' and its corresponding inertia in technological investment, cars with internal combustion engines are unlikely to disappear as the main means of transport any time soon. This will mean that it is likely we will still need to produce liquid fuels such as diesel fuel oil into the second half of the century (Dennis and Urry, 2009).

The increasing demand for fuel oils has led to a large increase in the agricultural production of fuel oil (biofuels), such as rapeseed oil or palm oil. Although renewable, they cannot be seen as sustainable, as they take land away from food production or encourage deforestation on a huge scale. The scale of production necessary for biofuels can be simply illustrated: taking Liverpool as an example, just meeting current energy demands for the city through bio-fuels would utilize an area of land larger than the county of Lancashire (see Figure 2.2). This is obviously unfeasible and unsustainable.

If Liverpool is to thrive in the next century, then a new way of operation is

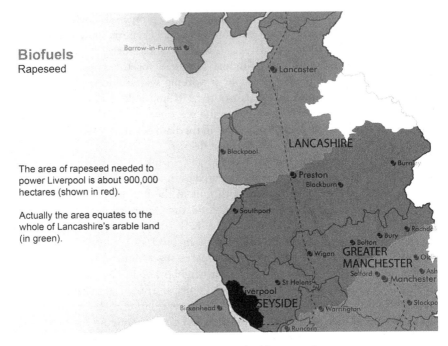

Figure 2.2 Impact of biofuels: powering Liverpool with rapeseed

needed that utilizes its connection with the sea, and that also improves its precarious position with regard to energy. One such solution may be to embrace large-scale algae production.

Algae culture offers another method of producing biofuels on a large scale. Algae are an aquatic plant that does not produce woody growth. Instead, it creates structural rigidity by filling its structures with oil. Some species of algae are up to 45 per cent by weight vegetable oil (Briggs, 2004). This vegetable oil is easily extractable by pressing, and is not dissimilar to diesel oils in fraction, but is much cleaner, having and exceptionally low sulphur content. Algae not only has a high hydrocarbon content, it also has a high productivity due to its almost continuous production cycle: through continuously filtering the mature algae out of the water, light penetrates into the water encouraging more rapid growth. This allows a much more efficient production cycle than land-based crops and thus a much higher yield. Algae species such as spirulina and *Botryococcus Braunii* have a high lipid content and fast growth and are the most suitable for UK systems as they grow in colder environments.

Hardware

The system that grows algae is known as an algae reactor. There are several available technologies for these, including open tanks, tubular systems and glass topped raceways. Closed raceway systems are slightly more expensive to build than open tanks, but being closed to air offer less chance of cross-fertilization and contamination with other less productive strains, and also reduce evaporation losses. Their other benefit, utilized in this project, is that they also facilitate forced carbon dioxide sequestration. Due to the UK's rather low temperatures in winter, these closed tanks are the most appropriate as the greenhouse effect keeps the tanks warm and thus productive (albeit at a lower output) throughout the year. Research has shown that this sort of system could theoretically produce around 150,000 litres of biodiesel per hectare per annum. This is a hundred times greater than a similar area of rapeseed (National Renewable Energy Laboratory, 1998). New technologies such as light-gathering rod technologies (to make light penetrate deeper into the tank) and carbon dioxide diffusers (which make CO_2 better available) could increase the yield by two or three times. The challenges the technology faces are mainly reducing maintenance and methods of control of the system. However the main problem is in the control of cross-contamination with lower-yield species, which could reduce output considerably.

Operation

The closed raceway system is simple in construction and operation. Water is circulated by a paddle pump around a circular bed approximately 500 mm deep. The pump has low power consumption and the paddle is designed to aerate the water as it circulates. Algae are continually collected as the water passes through a micro filter. In some systems a centrifuge is used to extract the algae. The system needs nutrients to be added to keep it productive. These are carbon dioxide, which is usually produced by a cogeneration plant (which may also supply low-grade heat to the tank in winter), along with phosphates and other minerals. These are usually obtained from composted sewage sludge or manure.

Process of implementation

The Bio-port Project was conceived as an emergent process-based urban transformation, with relatively low startup costs. Many future visions for cities are conceived without a roadmap to the final form and operation. Here the process of implementation was considered key in developing a future vision. Thus the project can be considered as 'realistic' and possible: none of the technologies are novel or untried. In Bio-port, it is the scale of implementation and the closed cycle urbanism that accompanies it, that makes the project innovative. The first investment is in a glass recycling plant. Europe has a large waste stream of powdered glass that is readily available, and once recycled this is the major component of the closed raceway tanks. The raceway tanks themselves can be floated in the Mersey Estuary. This offers several advantages: first it allows space for the development of a large-scale array, without disruption to the city; second it provides a relatively compact solution, occupying space between the major centres of population, namely Liverpool and Birkenhead, which minimizes supply distances for oil and heat and power; and finally it offers the chance of building a barrage across the mouth of the estuary to help prevent tidal inundation of the city from storm surges in the Irish Sea.

The process of implementation is simple: glass for recycling will be transported to Liverpool by sea and processed in the recycling plant. The plant will be powered initially by fossil fuels until the first algae reactors are functioning, then algae will power the plant, and then as more arrays come on stream, then algae will start to power the city. Once revenue arrives, investment will be made in a second glass factory to increase the rate of production, until ten modular glass recycling plants are in operation, which will, over a 50-year period, develop an energy production system that fills the whole estuary, a total of 6,000 hectares of algae array. This massive array will produce over double Liverpool's current energy demand for electricity and by a factor four over-supply heat, making Liverpool not only sustainable but also a net exporter of oil.

Software

In the HSI model described in this chapter, software is the biotic component of the system. There are literally countless species of algae available, however not all are suitable for biofuel production. Those that are need to have high lipid content, a high growth rate and must fit with the climatic conditions prevailing. Liverpool, at a latitude 53N, has a mild maritime climate, with an average temperature of around 5 degrees Celsius in winter and 15 degrees in summer. Algae species must be chosen to reflect these environmental conditions, bearing in mind those most suitable for Western Europe include spiro and *Braunii*, and these both require a reasonably low light level (around 60w/m^2) and low ambient temperatures.

Interface

The interface with the city for a productive system of this size is complex. First, the system lends itself to large-scale cogeneration, mainly because the fuel oil produced is easily transported and stored in the city. The current heating system for the city is based on natural gas, and a future district heating system could make use of the same infrastructure by installing evacuated piping directly in place of the gas piping. The fuel system for heat and power is just a part of the productive cycle. As approximately 50 per cent of the algae by weight is oil, there will be a large waste component of mainly cellulose, produced as a byproduct from the extraction process. The use of this cellulose allows resource cycles in the city to be closed. This connection to other productive streams is the 'interface' of the productive system.

In Bio-port the interface is directly with the food production system. The crushed algae waste is used as feed for cows in the hinterland of the city. This helps to maximize the productive potential of the region. Indeed there is further cycling, where the waste produced by the cows is utilized to develop market garden-based agriculture within the city. This circularity can be seen clearly in Figure 2.3.

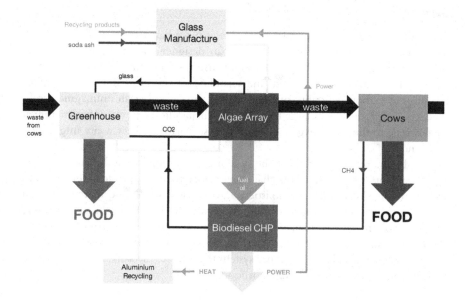

Figure 2.3 Bio-port resource flows

This urban market garden system has two functions: not only does it feed the city, but also by utilizing derelict land within the curtilage of the city, it helps to reconfigure the complex and broken urban spatial matrix that shrinkage has created. The resource cycle is finally closed when green waste from the greenhouses and human sewage are released back into the algae reactors to replenish the resource cycle.

Figure 2.4 Bio-port climax: after 40 years of growth, Bio-port fills the entire estuary and provides more than enough fuel for the whole city

Summary

Bio-port, although a theoretical proposition, is place based and realistic. It offers a view of the sort of interventions that urban designers can make in the city to increase resilience, without large-scale rapid change of investment. These emergent systems, built on a process-based understanding of the city, allow for a gradual adaption to carbon neutrality of the city over time, albeit at an unimagined scale. The scale of production is large but necessary: bringing the power infrastructure at such a scale to the city inevitably creates new scenarios for how a city might look and function.

The HSI approach shows that the system is only complete and useful when all three elements are present. This challenges urban planners and designers to work in a more interdisciplinary and holistic way, with the gain that the approach creates synergies to urban function that would otherwise be lost. Here the re-imagining of Liverpool creates a sustainable energy and food future without really affecting the functionality or form of the city. The next section of the chapter describes a much smaller scale urban agriculture system that utilizes the same methodology to deliver change in a deprived neighborhood.

Biospheric Project, Salford, UK: hyper-localized systems

The development of the globalized food system has stretched our resource net to the limit. In reaction to this there is a move toward highly tailored, hyper-localized systems based around localism, community growing and the utilization of 'unclaimed' post-industrial space. Critics of these types of systems cite that fact that subsistence-scale agriculture is not as efficient as its industrial counterpoint and that its output makes little difference to the consumption of the city as a whole. This critique misses the point of participatory urban agriculture: hyper-localized food production may not be efficient compared to industrialized farming, but it is extremely effective. It brings people together, educates people in health matters, greens the city, produces healthy crops, makes economic activity in areas of deprivation, can alleviate food poverty and reconfigure the city, helping to make it more legible.

The Biospheric Project

One such scheme is the Biospheric Project at the Biospheric Foundation in Blackfriars, Salford, Greater Manchester. The Biospheric Foundation was created by Vincent Walsh to readdress issues of sustainability and community in a deprived area of the inner city (Biospheroc Foundation, 2014). Blackfriars is one of the most deprived neighborhoods in the UK. Despite being situate within a kilometre of the city centre of Manchester, global wealth has bypassed the neighbourhood. A study by the Biospheric Foundation found the area to be a 'food desert', a place where it was almost impossible to buy fresh food (Hall, 2013). The Biospheric Foundation set up a project to change this, and through funding from the biennial

Manchester International Festival (MIF) created the Biospheric Project (Biospheric Project, 2014).

The aim of the Biospheric Project was to show that hyper-localized food systems could change people's lives. Interestingly, MIF had noted a connection between poor diet and a lack of connection with art, so it was their aim to link food directly with the arts festival. Working with researchers at Queens University Belfast, the project created an urban strategy plan to close cycles with respect to food in the neighbourhood, and the Biospheric Project would be central to this.

Issues with land

Early studies of the urban space in Blackfriars revealed several issues, which were related directly to post-industrial shrinkage. Population density was low, and the reconfiguration of the space after industry had left had created a patchwork infra-structure, with a large amount of unprogrammed space, which made the neighbourhood difficult to read geographically, with pockets of occupied land in isolated positions. Initially the team viewed traditional urban agriculture as a method for the reconnection of these spaces, but this was not possible. Like so much post-industrial space, all the available vacant land was highly polluted and unsuitable for growing crops. There needed to be another solution to the development of Blackfriars as a productive neighborhood.

The solution developed was two-fold: first soilless systems would be used; these could occupy buildings and land without engaging with the pollution beneath, and second, whilst these systems were in operation, phyto-remediation would be allowed to take place beneath the systems on waste ground, so that over time the land could be returned to a natural condition, and soil-based agriculture could be introduced at a later date. This long-term plan had other barriers to its implementation, both in scale of investment and expertise in particular as the team had little knowledge of soilless systems, and particularly their performance. It would be necessary to build a prototype system that could act as demonstrator for the bigger idea, so the Biospheric Project was created.

The Biospheric Project consisted of three components: a building-based technical food system; a community forest garden; and a prototype food network consisting of a wholefood shop, food box distribution system and hyper-localized connection with food industries.

Hardware: technical food systems

The technical food system developed was based within Irwell House, the home of the Biospheric Foundation. Developed with monies from MIF, it was intended to be part exhibition, part prototype and functioning laboratory. The brief was to develop a system that would inspire the public, but also leave a legacy, both as a productive system and also a testing facility for urban aquaponics.

Aquaponics are a combination of aquaculture and hydroponics. Aquaculture is the breeding of fish, whereas hydroponic systems are soilless growing systems where the plants are grown in water. The advantage of this technology is that nutrients are readily available and growth rates much higher than soil-based systems. Pests and disease can also be reduced. There are also some disadvantages, namely that energy needs to be expended to pump the water around the system, and that nutrients need to be added to the system, of which many are produced with fossil fuels or through traditional agriculture. In addition, as the systems have been developed as part of the intensive agriculture industry, they usually configured as monocultures and are not particularly resilient to crop failure.

The design brief for the project was a complex: the food system had to be designed to function not only as a laboratory for testing technically based urban agriculture, and also be profitable in use for the Biospheric Foundation, but in addition it had to be an appropriate exhibit for MIF, which was funding the project. MIF is a biennial arts festival, which aims to engage residents from all backgrounds in the city with creative endeavours. There are links between (poor) diet and (lack of) engagement with art, which the festival was keen to address.

System description

The design of the system was a compromise between the conflicting needs of a project that had to be efficient as a production facility, able to be analyzed as a laboratory and finally be an excellent visual exhibition. On these counts, the decision was taken to install the system partly within the building and partly on the roof, where light levels are high. The more visually exciting parts of the system were contained on the second floor of the building, these were the fish themselves, the mineralization bank and the deep-rooted crop bags placed in the south-facing windows. The system on the roof consisted of the nutrient film system for growing leaf crops, and this was contained within a polytunnel.

The system is relatively simple in design (see Figure 2.5): there are 12 fish tanks which are fed with returning water from the nutrient film (NFT) system on the roof, the overflow from the fish tanks collects in a sump, and the water from here is pumped to the mineralization bank. The water here drains consecutively through a series of syphonic containers containing expanded clay balls and worms into a further sump, from where it is pumped across the ceiling of the old mill and drains through silicon bags hanging in the windows where fruiting plants such as tomatoes are grown, and from here it is pumped up to the roof into the polytunnel, where it flows down through 40 nutrient film channels, containing over 6,500 leaf crops, and back to the fish tanks. The risk of legionella build-up in the system was considered low but as a precaution, each time the water was pumped, it passed through a UV filter.

The chemistry of aquaponics is rather complicated, but a simple description is that the fish are fed (in our case with organic fish food) and they produce two types of waste: ammonia through their gills and solid waste, both of which are toxic to the fish and, in the form they are in, their nutrient content is not available to plants. This is the function of the mineralization bank: here nitrating bacteria transform the ammonia first into nitrite and then into nitrate, which are available nutrients

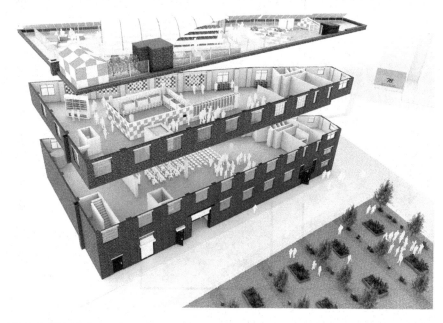

Figure 2.5 Exploded axonometric of the design showing second floor aquaculture lab and roof-top greenhouse

for the plants, and the solid waste is processed by the worms into nitrate. Many systems filter out and dispose of the solid waste, rather than process it, but in the Biospheric Project it was thought that this was against the ethos of the closed cycle philosophy of the project.

The sizing of the system is related to the chemistry of the process, and can be difficult to optimize. In the case of the Biospheric Project it was decided to use the method described by Wilson Lennard (Lennard, 2012). Input and output for theses sort of systems have not been fully researched, and this was to be part of the project's brief as a 'living laboratory'. In the Biospheric system, the there were 12m³ fish tanks, 90 400x500x200mm mineralization beds, 130 window growing bags and a 6,500 plant NFT array.

Technical issues

The novelty of the system created some design issues, particularly related to the age and condition of the near-derelict building. These were related mainly to the weight of the system. Soilless systems are heavy, and existing buildings are not designed to carry the extra weight of the water. The fish tanks alone weighed over 12 tons, and the building floor could not take the weight. The tanks were arranged in a square plan along major beams to reduce deflection. Similarly the mineralization bank was

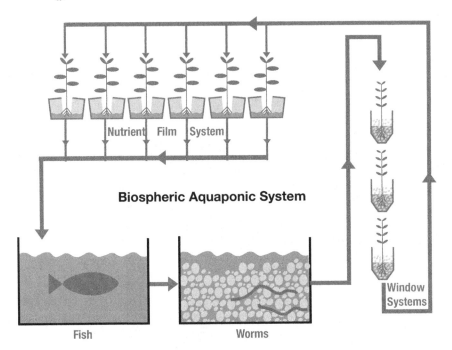

Figure 2.6 Aquaponic process diagram

positioned along a structural beam. The roof was also not designed to carry the load of the polytunnel greenhouse or the added weight of the NFT system. To solve this a series of structural adaptations were made consisting of two additional beams that spanned existing structural columns, on which the polytunnel was mounted. The second issue related to this was the need for a sophisticated control system. The structural issues of the building meant that the sumps had to be kept separate. Thus the three pumps had to work together to balance the system, and prevent one part of the system running dry. A series of flow sensors and float switches was needed.

Software: permaculture and biodiversity

Seeing the biotic as a separate entity from the hardware of the system allowed the team to tailor the system once designed to suit the local context: climatic, social and economic. The nature of the project and the clientele meant that the software of the system was designed around ideas of permaculture and biodiversity. Many aquaponic systems are monocultural in design – growing one vegetable crop and one fish, and utilize chemical or adulterated feed. The idea of the Biospheric Project was to be organic and as biodiverse as possible as the raison d'être of the Foundation was to develop a more biomimetic approach to agricultural production.

Figure 2.7 Aquaculture as exhibition (clockwise from top): glazed fish tank; mineralization bank with cover removed; window system; window system detail

Figure 2.8 Roof-mounted polytunnel with nutrient film system

Note: Note the bio-diverse planting.

The total number of crops in the system at any one time is over 6,500. These are mainly leaf crops, together with a smaller number of fruiting crops. Over 30 different varieties and species were grown at one time, including kale, Good King Henry, sorrel, tomatoes and peppers. Plants were chosen according to the ideas of companion planting in permaculture (Choose Permaculture, 2014). This method of planting combines crops with other crops that need differing nutrients and have resistance to different pests; the plants support each other and should reduce the need for pest control. The plants chosen were also generally indigenous, as it was thought this would help with increase biodiversity as well as pest control.

As well as the plant crops, the system was built to contain four species of fish, in 12 tanks, suitable for three sizes of each species. To commission the project two species were chosen: Nile tilapia and carp. The reason for this was that both species are very resilient to water quality fluctuations and thus were more suitable for an experimental system. In the future, the project hopes to cultivate cold-water species such as trout or perch. Over the first year, the system produced over 100kg of fish.

Interface

Economic interface

Urban agriculture on this scale needs a carefully designed interface with consumers, both professional and local. The Biospheric Project had this interface from the start, both with the local food businesses and with local consumers, which cut across both ends of the food spectrum. The project worked closely with a highly rated restaurant in Manchester city centre, one that specializes in cooking with locally grown produce. An agreement was put in place for the restaurant to take the crop each day, at a fair market price, as long as the quality matched that from professional growers. This hyper-localized connection meant that freshly picked leaves could be in the kitchen with an hour of harvesting, and be transported directly by person or bicycle. This drastically reduced the food miles in salad crops and provided a service that could not be matched. The speed of turn-around also allowed negotiation directly with the restaurant on the type of crops that they wanted and when they wanted them, which minimized waste and maximized profit.

In addition to this, leaves that were imperfect or unrequired were sold to the public at the 78 Steps Wholefood shop nearby, which is part of the Biospheric Foundation and named after the 'food miles' from the system to the shop. The system's first crop in August 2013 produced 21kg of mixed leaves. The team who harvested the crops was amazed at the rate of growth (which is faster than soil-based growing) and the quality of the plants (Biospheric Foundation Blog, 2013). Further research on a wider scale is needed into the performance of systems such as this, but the system performed at the output predicted by the team.

Ecological interface

The Biospheric Project is the first part of a more integrated urban hyper-localized food system. Aquaponic systems, as efficient as they are, still need input in terms of fish food and other nutrients, and these need to be sourced sustainably. Typically 1kg of fully grown tilapia need around 10g of fish food per day, which should be around 35 per cent protein. This can be difficult to source sustainably, as most fish food is made of fish protein, which seems ridiculous. In the closed cycle strategy plan for the Biospheric Project it was envisaged that the protein would come from worms from a vermaponic waste food composting system, and from spent grain from a microbrewery. At present, these sources are not in great enough in quantity to feed the 150kg of fish in the current system, and their diet is being supplemented by organic fish food. Work continues in trying to close the material cycles regarding food in the neighbourhood.

Social interface

The project has attracted a great deal of attention, both locally and internationally, and the Biospheric Foundation has been inundated with volunteers and visitors. It is obvious that there is great interest in locally produced food, and consequentially eating healthily. The aquaponic system, although a technical system with controls and pumps, is still farming, and is labour intensive. Seeds need to be cultivated, plants harvested and pests need to be controlled. This all takes man-power, but of a rather low-skilled variety. This can only be a good thing, particularly in localities in neighbourhoods such as Blackfriars, where the sense of community is weak, unemployment is high and the diet poor.

Conclusion

The model for urban agriculture developed in this chapter is an integrated and systematic one, where there are not only biotic components, but also technological and sociological ones. Although this approach allows the design of the system to be considered as urban design, and its operation to be considered in another way, it must still be seen as holistic. All parts of the system of hardware, software and interface still need to be considered in order to make a fully functioning urban system, but their separation allows designers to make decisions that are not bound by any one element in detail. In other words, this systemization may make the design process simpler but it does not help the implementation and operation of these types of systems. This merely confirms and reinforces the fact that the design of urban agriculture is more complex than most urban design interventions because it deals with disparate subjects from urban morphology, agricultural technology, plant science and the complexities of urban food networks. Beyond this even there are issues with land use in cities and the scale of project needed to make a difference, which will have to be addressed if we are to make cities more resilient in the future.

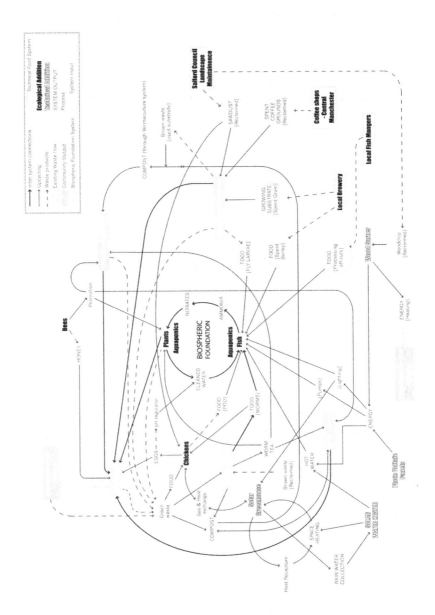

Figure 2.9 Closing cycles in Blackfriars

The Bio-port Project, although propositional, utilizes technologies available today, but at a scale of magnitude previously unimagined. It would take a huge political will to deliver such a project, even though the initial investment would be quite low and the pathway to implementation simple. The hardware/software/ interface is a clear analogy here, and it is the interface part that is the most interesting. Here the interface is not only with the technological cycles of energy infrastructure through the recycling of glass and aluminum and district heating and smart-grid technologies in the city, but also with the local biotic and agricultural systems, through feed for cattle and urban farming green waste. The complexity of the system and the need for holism cannot be underestimated, and would need to be developed concurrently, which is always problematic for urban projects. What is sure is that if we are to move to a new biofueled era, then there will need to be a redefining of the governance of the city, and a re-imagination of the land use of its hinterland.

The hyper-localized implementation of soilless agriculture also has a clear adherence to the HSI proposition, but here the issues are different. In technological terms, the systems, integrated into the existing fabric, reveal structural and control issues, which need to be factored into the mass adoption of these technologies. In addition, the complexity of the hyper-localized food network of consumers and producers is essential for the system to be functional, but needs local knowledge and willing participants. Further complexity in this project was the implementation of community ownership and operation. The system has not performed as well as previously since hand-over, which is mainly due to the amount of input needed on a day-to-day basis to ensure correct operation of what is a highly complex and novel technological biotic hybrid. It seems that despite the technical acuity of soilless systems, the agricultural knowledge and input needed to keep the systems operating productively is still high, and is as crucial a part of the system as the other components.

To conclude, both projects show that technological urban agriculture is still in its infancy, but there is potential that its rapid and whole-scale implementation could dramatically change cities, not only making them more resilient and sustainable, but also in helping to bring together communities, particularly those under economic pressure. However, the complexity shown by the analogy of the HSI model, which covers the whole spectrum of urbanism, will need to be taken into account if this implementation is to be effective. This is an issue not only in the design of a system or its implementation, but perhaps more importantly in the way the systems are embedded into the lifestyles of operators and consumers of the products of the system. The relationship between food and the city is complex and the reconfiguration of this relationship has the potential to be a powerful and revolutionary force for change in the urban condition.

Acknowledgements

The author would like to acknowledge the hard work and dedication of the following in the design and production of the projects described in this chapter:

Simon Swietochowski, Andrew Jenkins, Tilly Hall, Vincent Walsh, Morgan Grennan and Josh Greenfield.

References

Abt, J. (2011) *American Egyptologist: the life of James Henry Breasted and the creation of his Oriental Institute*. Chicago: University of Chicago Press.

Appadurai, A. (1996) *Modernity at Large: Cultural dimensions of Globalisation*. Minnesota: University of Minnesota Press.

BBC-News (2000) Refinery hit by fuel protesters. http://news.bbc.co.uk/1/hi/uk/915567.stm. 8 September. Retrieved 12 January 2008.

BBC-News (2014) Ukraine crisis. www.bbc.co.uk/news/world-europe-27804611. Retrieved 8 May 2014.

Best Foot Forward *et al.* (2005) *City Limits: a resource flow and ecological footprint analysis for Greater London*. www.citylimitslondon.com. Retrieved 10 March 2014.

Biospheroc Foundation (2014) Homepage. www.biosphericfoundation.com. Retrieved 10 May 2014.

Biospheric Foundation Blog (2013) Biospheric first crop. www.biosphericfoundation.com/biospheric-first-crop. Retrieved 5 September 2013.

Biospheric Project (2014) Urban Farm Laboratory, Research Centre. www.mif.co.uk/event/the-biospheric-project. Retrieved 10 May 2014.

Boycott, R. (2008) Nine meals from Anarchy. How the UK is facing a very real food crisis. *Daily Mail*. 7 June. www.dailymail.co.uk/news/article-1024833/Nine-meals-anarchy—Britain-facing-real-food-crisis.html. Retrieved 10 June 2014.

Briggs, M. (2004) *Wide Scale Biodiesel Production from Algae*. University of New Hampshire, Physics Department: UNH Biodiesel Group.

Choose Permaculture (2014) Companion planting. www.choosepermaculture.com/companionplanting.php Retrieved 10 May 2014.

Dennis, K. and Urry, J. (2009) *After the Car*. London: Wiley.

Gundlach, R. (2015) *Greg Keeffe – Why urban food production makes sense and how to make it happen*. International Urban Food Network, 29 January.

Hall, N. (2013) Blackfriars resource analysis. Belfast: Architecture at Queens internal report.

Lennard, W. (2012) Aquaponic system design parameters. Aquaponic Systems. www.aquaponic.com.au/Fish%20tank%20shape%20and%20design.pdf.

National Renewable Energy Laboratory (1998) *A Look Back at the US Department of Energy*. Washington DC: NREL.

Oswalt, P. (ed.) (2005) *Shrinking Cities V1*. New York: Hatje Cantz.

Paull, J. (2013) 'Please Pick Me' – How incredible edible Todmorden is repurposing the commons for open source food and agricultural biodiversity. In J. Franzo, D. Hunter, T. Borelli and F. Mattei (eds) *Diversifying Foods and Diets: Using Agricultural Biodiversity to Improve Nutrition and Health*. Oxford: Earthscan, Routledge.

Rickman, J.C., Barrett, D.M. and Bruhn, C.M. (2007) Nutritional comparison of fresh, frozen and canned fruits and vegetables. Part 1. Vitamins C and B and phenolic compounds. *Journal of the Science of Food and Agriculture*, 87(6): 930–944.

Roberts, S. (2010) *The Oil Crunch: a wake-up call for the UK Economy*. London: Arups.

Roggema, R. (2014) FoodRoof Rio: How favela residents grow their own food. www.adjacentgovernment.co.uk/lg-edition-004/profile-foodroof-rio-favela-residents-grow-food/11807/ Retrieved 25 September 2015.

Roggema, R., A. Pugliese, B. Broekhuis, and M. Drissen (2014) The FoodRoof: How Cantagalo and Pavão-Pavãozinho favelas grow their own food. In: Roggema, R. and G. Keeffe (eds) *Why We Need Small Cows: Ways to Design for Urban Agriculture*. Velp: VHL University Press.

Save Liverpool Docks (2008) *Idle Land*. Liverpool: Save Liverpool Docks.

3 The cultural landscape of food

The infrastructure resilience of Via Emilia

Anna Chiara Leardini and Stefano Serventi

Italy is well known for its gastronomic resources, as is its strongest agricultural basin: one of the most productive landscapes is the territory represented by Emilia-Romagna (Gallia Cispadania, the southern region of the Po Valley), reflected in its main infrastructure system: Via Emilia (also known Via Æmilia). The aim of this chapter is to determine the current conditions of the rural territory in the region and problems affecting the current socioeconomic assets in order to propose a strategy to rehabilitate the physical infrastructure of Via Emilia as a cultural landscape and food heritage site. The chapter suggests that considering the whole territory and the pattern of its agricultural plots (Roman centuriation in this case), and thus the rural grid related to its geographical characteristics, is an important basis for developing planning strategies, introducing the concept of *foodshed* as the indicator of analysis.

An increasing number of design and research projects in landscape urbanism address issues of food and its relationship to landscape. The production, distribution, processing, marketing, consumption, and disposal of food have consequences for landscape processes and cultures. This chapter articulates how ecological, cultural, and spatial systems have mutually influenced food and productive landscapes and vice versa, creating what we call a cultural landscape in the Emilia-Romagna region.

Since the Roman Empire, Via Emilia has been the head of a complex territorial organization system, in which the whole region can be considered as a single city-structure. Caused by the first ecological revolution, this territory emerged as organized without a division between urban and rural areas. Here, rurality corresponds to a certain approach to colonizing and conquering the environment using agriculture.

Centuriation indicates the fundamental structure for the organization of the territory. It is therefore understandable that the importance of the centurial "net" maintained its importance even in the post-Roman era, today resulting in the concept of rurality being considered as a diffuse organization.

We believe that food production here in the region is represented not just by its intrinsic quality, but also by its placement as a result of the intricate mythology of territorial organization. This cultural production represents the relationship between man and his environment (Gambi, 1950).

We consider this infrastructure a cultural element; it is the intricate element that made the Italian valley capable of producing food over the course of centuries, in turn becoming the most important Italian gastronomic culture, but also able to bring together strategies for rural development and spatial design: a model of cultural heritage diffusion.

After a review of the data on the current and future perspectives underlined by the Rural Development Program, the term *territory* has to be analysed firstly as a way of land occupancy for military or civil aims, but also as a result of the Roman colonization that has characterized the history of the Emilia-Romagna landscape. Research on these issues, imagining a scenario for a rural development, could represent the possible shift in which the intervention's scale does not refer to a series of places but to the whole *territory*.

Subsequently, the territory of Emilia-Romagna will be reviewed in its Roman configuration, the *Centuriatio*, as a rural model and moreover as an ecological structure. This system strongly transformed the environment, allowing the development of a productive landscape based primarily on agriculture and husbandry land use.

After having analysed three possible structures (spatial, cultural, ecological), the complex infrastructure of Via Emilia as the head of the entire territorial system will be considered, and its context will be shifted to the ecological sciences in order to find a new operation that can generate a possible strategy for rural development and cultural heritage.

The concept of *foodshed* introduces potential strategies for shaping territories and landscapes historically defined as site-based projects: "It argues that designers should expand practice from an emphasis on local food strategies of productive spaces to respatialize the logistical spaces of aggregation and distribution in order make significant structural change in both landscape and food systems" (Potteiger, 2013).

Defining territory

After the expulsion of the Gauls, the Romans began the colonization of Emilia-Romagna (Gallia Cispadania) with the great construction of Via Emilia. The construction of this consular road, 176 miles long, was part of a series of major infrastructure projects that would have reconfigured the geopolitical boundaries of the Italian peninsula. The role of this *decumano* was not in fact just to connect the centre of Italy to the Adriatic Sea, but rather to colonize an area that until then would not allow the establishment of agriculturally productive activities. The Romans acted on a new scale that was indispensable for registering and engaging the complexity of networks and information in a given physical environment. This means that the idea of territory at that time included a series of social, political, and economic factors that engaged the extrinsic environment of the place. When we think about spatial configuration, we are looking to shift the idea from acting on a particular site to acting on a territory, which includes global and local factors.

In French, the noun *terre* comes from the Latin verb *terra* (dry land, as opposed to the sea; earth) and has two different meanings: *terroir* is a geographic area,

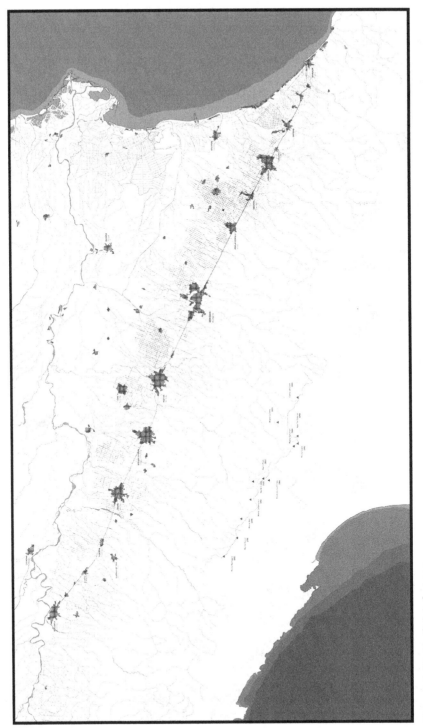

Figure 3.1 Via Emilia's infrastructure

homogeneous in terms of resources and production, often related to agricultural production and tied to distinct cultural communities. For Barham, it also refers to "an area or terrain, usually rather small, whose soil and microclimate impart distinctive qualities to food products" (Barham, 2003). However, a *territoire* is a political or jurisdictional region, often defined by a border or natural territory, defined by more ambiguous ecological delimitations.

As we see from this diversification in the complex Roman colonization, the territory is something that gathers together all of these characteristics. The military idea of conquering the world and building the strongest and most extended Empire supported the notion we have of a territory, which becomes more complex when considering the conditions of globalization – the spaces of information exchange, of international trade and retail, or of military occupation.

The Emilia-Romagna region is an individual site for agro-alimentary production that serves multiple trade networks all around the world, determining deep flows of information and thus influencing the technological and ecological environments merging into ambiguously delimitated boundaries, operating at multiple scales.

A possible way of analysing this territory at the micro and macro scales is represented by one of the main elements of Roman conquest, which in the design of productive patterns was adopted to control and divert natural resources. This element, called *Centuriatio* in Latin, is a particular organization of the territory created by the Romans after their conquest, and it responds not only to productive functions but also to the Roman vision of life and the cosmos in a religious

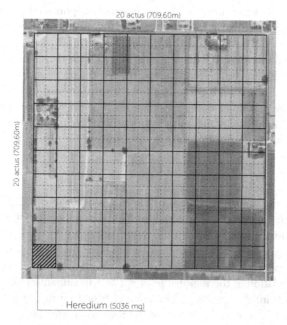

Figure 3.2 Palimpsest of the centurial grid

sense. The first land divisions had the aim of creating an extensive and coordinated defensive belt against external attacks, a belt formed by the houses of the settlers that were built on individual properties and scattered throughout the colonized agricultural areas.

There were two types of centuriation: *secundum caelum* (according to the heavens), in which the priests translated the organization of the celestial vault onto the land (this is an adaptation by Etruscan culture), and *secundum naturam* (according to nature), in which the centuriation was adapted to the soil according to the physical characteristics of the territory. The division and allocation of the *centurias* were carried out according to the military scheme of the camp: the *limitatio* perhaps derived from the installation of the rectangular *castrum* (military camp), which emphasized the rational Hellenistic scheme (the urban *Hippodamian* pattern).

Centuratio was the main system of parcelling land and it was extensively applied. It was essentially a regular grid consisting of squares of about 710 meters per side (corresponding to 20 *actus*, 1 *actus* being approximately 35.5m) and materialized on the ground in the *cardo* (oriented approximately north–south) and *decumano* (oriented approximately east–west), which in turn coincided with connecting roads and ditches. A *centuria* of this type corresponded to 200 acres (close to 50 hectares), and was further divided into 100 heredia, each of which consisted of two *iugero* (*iugerum*) a unit of surface area equivalent to 0.25 hectares (Bazzani, 2004).

The dominion was not formed only by the limits (*cardo* and *decumano*), but also by a dense array of subdivisions within individual *centurias*: these cultural ditches and drains, as well as the "headland", gave material support to the borders of various kinds, together with the planting of trees (elms) and a particular grapevine type.

So a real system emerged, which was entrusted with the task of ensuring road links between the various parts of the territory, drainage, and irrigation of the countryside, and indicating the backbone of the territory and the cadastral boundaries of land ownership: a complex organism that only a state apparatus could promote as highly efficient and, most importantly, which was maintained through the assiduous care provided by local populations.

Data

As reported in the June 2015 update of the Emilia-Romagna Rural Development Plan, the region has an area of 22,445.5 km² and a population of 4,342,135 inhabitants. The local institutional framework includes: 9 provinces and 341 municipalities gathered in 40 municipal associations. In the region, according to Eurostat, 76.7 per cent of the total population resides in rural areas. The average density of population at the regional level is 198.1 inhabitants/km², while in rural areas it is 124.6 inhabitants/km². The areas classified by region as "less favoured mountain zones" cover 36.7 per cent of the region's surface area.

At the regional level, the classification of the degree of rurality defined by

Eurostat was refined to outline more accurately those territories with greater social and economic problems, in accordance with the national methodology. As a result, the following types of rural areas were identified:

- Urban poles (zone A)
- Areas of specialized agriculture (zone B)
- Intermediate rural areas (zone C)
- Rural areas with development problems (zone D)

In areas marked with a greater degree of rurality (zones C and D), we find 69 per cent of the regional area and 33 per cent of the total population.

The Rural Development Program 2014–2020 of the Emilia-Romagna Region, promotes a new strategy to align the agro-alimentary sector with the *Guidelines for the Ex Ante Evaluations of 2014–2020 of EU 2020 Strategy.*

In SWOT analysis, the socio-economic structure of the Emilia-Romagna region is characterized by a position of relative socio-economic advantage based on income levels (ICC 8) and the general welfare of the region. Like all its provinces, is at the higher end of the field (EU27). However, like the whole of society and the national economy, this region is experiencing its deepest crisis since World War II, not just because it is suffering from a long period of low economic growth, but also due to clear signs of a real decline. In light of this context, some indicators of general scope should be noted, which, with respect to the perspective of the EU 2020 Strategy, indicate some problems in the evolution of regional society and its economic development pattern.

Long-term processes that are profoundly changing the characteristics and needs of local communities, in particular aging and immigration, lead to at least a partial overlap in negative conjectural trends: the employment rate fell to 67.6 per cent, while the unemployment rate reached 7.1 per cent and 26.4 per cent in the young, and poverty affected 14.9 per cent of the population as of 2011.

Although the labour dynamics are better than the national average (72 per cent vs. 61 per cent employed in 2010), rural areas show lower rates of employment. Wealth (GDP per capita) is lower than the overall regional figures, and the internal rate of the population at risk of poverty or social exclusion has reached 14.9 per cent, while the degree of relative poverty of households (ISTAT), although less unfavourable (5.2 per cent), still shows a strong negative trend (Regione Emilia-Romagna, 2015).

Despite the framework described above, the food system has always been a productive sector at the regional level, characterized by a marked distinctiveness and an indisputable success in global competitiveness based on quality; at a critical time, it shows itself able to maintain and even increase its influence inside the regional production system, confirming its anti-cyclical and stabilizing nature, as well as representing one of the most important sectors of incremental change in the evolution of the regional economic model.

The production system in Emilia-Romagna is characterized by the importance of a number of agro-food products, whose dynamism has enabled the mitigation of the negative effects of the financial and economic crisis on the regional economy. The fruit and vegetable sector is the most important sector in the region, considering that 24 per cent of agricultural production in value are products from Emilia-Romagna, in particular those based on husbandry and dairy sectors, which also represent the excellence of the "Made in Italy" brand.

The success of the Emilia-Romagna region is to be found in particular in the quality of its agricultural food production. In Emilia-Romagna there were 33 registered Protected Designation of Origin (PDO) or Protected Geographical Indication (PGI) certifications, making Emilia-Romagna the most important food production region in the nation. However, the system of typical regional products has reached an advanced stage of maturity, which must be consolidated and enhanced through the expansion of signatories to quality designation and certification systems.

The agricultural sector suffers from obvious problems of competitiveness, as shown by the negative profitability of the invested capital in the agricultural sector, the high incidence of the production costs of sales and labour productivity, which while better than the Italian average, is lower than the average in the European Union. However, the processing industry is dynamic, competitive, growing, and drives the entire food system.

As reported in the new Rural Development Program (*Programma di sviluppo rurale 2014–2020*) of the Emilia-Romagna region, one of the main problems in the competitiveness of the agricultural sector is represented by the fragmentation in the productive phase, to which is opposed a concentration of processing and commercialization activities. The small size of production combined with the lack of bargaining power that results from it are the foundation of the strong imbalances in the supply chain, and thus also in terms of the distribution of value, to farmers' disadvantage. This is called *agricultural pulverization*, due to inefficiency and delay in logistic and commercial development.

Another serious problem is represented by the aging population. Farmers over 55 years of age represent 60 per cent of the sector that could retire without successors. This means that in the future, more than 50 per cent of the cultivable surface could be abandoned and with it the production for which the Emilia-Romagna region has been successfully competitive.

To counter this problem in the past, regional political leaders have reacted using different strategies, aggregating the small-size production chain horizontally (cooperatives, associations, and so on), and vertically (contracts, framework agreements, professional organizations).

With respect to this issue, our analysis of the territory and food systems seeks to rehabilitate the alignment and adaptability between the system and its environment, on which the performance of any organization depends. In doing so, our aim is to consider the productive sector of the economy, the agro-alimentary system, as an element that has to be taken into account in rural development strategies. It is extremely necessary to rethink the patterns, "the contents and internal order of a heterogeneous area of land" (Forman, 1986).

Figure 3.3 Centuriation resilience – an aerial view
Source: Courtesy of Eng. Arturo Colamussi.

Ecological structure

The first reasonably accurate recordings of the most prominent features that shape the region's landscape are found in writings dating from the conquest of the Roman Empire. Between the second century and the Imperial Age, Polibio, Strabone, Livio, and later Plinio defined the three main areas of the region: the fertile Po (south) Valley flowed through the eastern peninsular valleys, the foothill belt, and then the Apennines. The first zone was the perfect place to create a settlement utilizing the complex centuriation process, representing a longstanding desire to occupy the area because of its useful economic potential.

Since the Neolithic era, this area has been occupied by the most urban settlements in the entirety of Italy (or what was considered the political territory of Italy at that time), and thus also the presence of primordial agriculture (cereals) and livestock (sheep, goats, pigs, and oxen) (Gambi *et al.* 2008). In the differentiation of various agricultural structures, the Pianura Emiliano-Romagnola is characterized by the diversity of its crops: although the land appears in approximate regularity, in the shape of rectangles or squares, the originality and at the same time the main problem with this type of structure is to bring together in the same field, not two but three crops: those of herbaceous plants (cereals, forage crops, and rotating crops), a shrub (i.e. the grapevine), and trained or staked trees. It is not just a problem that affects genres of crop or production, but it is often a fact of agricultural structure in the strict sense of the word. Human settlement is usually missing: the farmers live in isolated houses or are gathered in tiny hamlets.

Because of its geographic localization, Roman colonists settled down in villages following their supply system and the use of resources; the land was radically transformed by intense deforestation and reclamation of wetlands, causing a reduction in biodiversity that still persists today (Mazzotti, 2003).

Rural model

The idea of organizing nature in order to create a liveable model of habitat between wetlands – that was the Pianura Padana (Po Valley) during the Roman Empire, and corresponded to what we call "ruralization" today, and respected the theory of Pierre George, a master of rural studies, who defined the rural landscape as a thoughtful and concerted action on the natural environment, "*une action réfléchie et concertée sur le milieu naturel*". In his view, "*la sédentarisation*" (settlement) implies a progressive and sustainable management of the occupied and especially the cultivated space, including permanent human oversight of various matters: river regulation, protection against floods, depletion of local water supplies for irrigation or for use in water powered mills for the processing of agricultural products; differential land use according to the abilities of each sector of land to respond to the type of farming required for specific activities; and establishment of essential routes for agricultural vehicular traffic and possibly trade. The landscape acquires a face that deviates more or less from the original appearance of the occupied areas. Each agricultural civilization's ability to transform the environment depends on the

Figure 3.4 Typical farm along the Via Emilia-Romagna

Source: Stefano Serventi.

relationship between the resistance of the environment to this transformation and the creative power of the farmers (George, 1956).

Rurality is linked to a condition where culture and production built a model of territorial organization, and the acknowledgment we have about Gallia Cisalpina is expressed in "the treatise on Roman agriculture". "*Apud Romanos bonus civis bonus colonus erat*" (with the Romans a good citizen was a farmer). The citizen, in fact, was not only the inhabitant of the city but was part of the community, with a specific number of duties and similar rights. The settler was the citizen of the countryside and therefore a farmer.

The rural model, as mentioned in the description of land division, was thus the idea that the military, social, and economic spheres avail themselves of the same rules of occupation of the territory. As documented in Roman history, this cosmic vision of territorial organization was regarded as the paradigm of all human constructiveness and therefore as culture (Bazzani, 2004).

Not surprisingly, the noun *cultus*, from the past participle of the verb *colère*, indicates not only "to cultivate" fields, "to grow" crops, but also to care for something in general and in the specific sense, as in the cultivation of human beings.

Emilia-Romagna is in major part defined as a rural landscape, and at the same time is the result of the spatial planning for agro-alimentary production that strongly influenced the environments, enhancing the possibility of introducing new social structures and new economies.

However, in the description of the rural landscape by the geographer Lucio Gambi, he rightly states that a more careful examination of these features are as parts of a much more significant complex, and in reality are tightly and inextricably linked to many human phenomena that do not remain reflected in the topography and as a result of events or institutions or human structures are only partially able to engage the senses (Gambi, 1961).

One of these phenomena is the relationship between the individual and the group: as is the case in a society where individuals are free in their acts and their techniques, and that is very different from a society where individuals are strongly tied to an institutional authority that exceeds them.

The example is the opposition in the rural history of the last six centuries: that is, the conflict between the egalitarian land distribution systems common in the Middle Ages and the capitalist selection that became manifest from the sixteenth century onward, which aims to stabilize the use of the fields for the individual and make them marketable or transferable by inheritance.

In the later centuries, rural development was concentrated on the intensive exploitation of land and natural resources through land use for agriculture and breeding. Intensive use and mass production are now in contrast, on the one hand, with changes in global production networks and urban growth, and on the other, with the idea of culture and rural development inferred from the ERND (European Network for Rural Development). Interest in landscape ecology, rural development, man's relationship with the environment, and so on is also expressed through writings such as the following:

The need for rural communities to approach development from a wider perspective has created more focus on a broad range of development goals rather than merely creating incentive for agricultural or resource based business. Education, entrepreneurship, physical infrastructure, and social infrastructure all play an important role in developing rural regions.

(Rowley *et al.*, 1996)

Understanding territory as a part of complex terrestrial ecosystems might be understood to produce a *re-territorialisation* – an anchoring of territory into locally specific, if dynamic, physical and ecological networks.

(Forman, 2005)

Rural land persists around the metropolis not because we have managed the land more wisely but because it is larger, more resistant to man's smear, more resilient. Nature regenerates faster in the country than in the city where the marks of men are well-nigh irreversible. Although we can still see great fat farms, their once deep soils, a geological resource, are thinner now, and we might well know that farming is another kind of mining, dissipating the substance of aeons of summers and multitudes of life.

(McHarg, 1995)

Figure 3.5 Via Emilia
Source: Stefano Serventi.

Infrastructure of resilience

With these principles, the structure of the Emilia-Romagna region is characterized by a single infrastructure, *the Via Emilia,* and thus seems to be the perfect place to rearticulate the discourse in the urban context, and even more for the sub-urban context. Its characteristics of being the main axis of the cities (so to say even different cities based on a single system) and the axis of the regional context can represent an ideological pretext through which to reformulate criteria, operational imperatives, and contemporary flows. As the Via Emilia was originally conceived, the main *decumanus* (which is actually the longest *decumanus maximum* in the world) was originally conceived as a way of locating and linking the cities along it and at the same time uniting the whole territory. This is can be an argument for rethinking how to integrate the urban and the rural zones into a new pattern. Since infrastructure can operate as a contingent ecology or managed dynamic system, this could be the spatial format in which contingent architecture might materialize.

This emphasis on the collective natural environment repositions the role of infrastructure as the foundational spatial format, as it allows for the interconnection between the human and environmental spheres, constantly negotiating the boundary between landscape, urbanism, and architecture. Infrastructure is implicated here once again as it oscillates between local and global as well as between the natural and artificial. In essence, infrastructure has been deployed as a machine over the larger landscape, which thrives from its subcomponents conforming to its logic.

While sub-networks of the infrastructural system offer a form of contingency, the essential conception and deployment has been to mitigate and eliminate disturbance. Although infrastructure supports almost all aspects of our daily lives – water, waste, mobility, energy, food, etc. – it has rarely been thought of holistically or symbiotically. The critical project that this chapter must undertake is a paradigmatic shift in the role and deployment of infrastructure, such that it operates symbiotically between the human and natural sphere as a soft system that provides resilience through the continual negotiation of "top down" and "bottom up" organizations within the ecological poles (Bhatia, 2013).

This understanding aligns itself with Felix Guattari's *The Three Ecologies*, wherein Guattari merges the human and environmental spheres into the concept of *ecosophy*. Guattari's *Three Ecologies* registers environment, social relations, and human subjectivities that could be re-characterized as the natural world and human politic (comprised of both the collective, or "social relations", and human subjectivities) (Guattari, 2000).

The ecological revolution has in fact transformed the landscape and "natural" elements are almost fictional; it is a complete altered ecology that has almost created its own biology from anthropological changes.

> The definition of ecology is both an organism's relationship to another organism, the "human ecology" (political, economic and social spheres), as well as an organism's relationship to its environment, i.e. "natural ecology" (including the design of landscape, infrastructures, urban form, as well as the

impact of environmental conditions such as geology, weather and ecosystems, for instance) ... As a decisive factor on how we approach the issue of economy, social integration, environment, and their associated spatial formats, the political sphere acts as a critical hinge in the reconciliation of the ecological project.

(Bhatia, 2012)

In the past, Emilia Romagna was strongly influenced by the human ecology that it had encountered through interventions in territory, the manner in which economics were controlled, and the effects of the social sphere on the environment. Today globalization, as the reconciliation of collective (equality) and individual (distinction), is a new human ecology "rooted in pluralism that can produce a collective agency, capable of restructuring our economic and socio-cultural territory and its relationship to the natural environment" (Arendt, 1958). This *continuum* slow alteration of the land introduced particular modern planning strategies with the intent of embracing different scales. According to a study conducted by Moussavi and Gareth, we cannot apply rules for the large scale as we did in the past in the planning of large-scale systems; a new alternative design approach is needed (Moussavi & Gareth, 2010).

In the future of a post-industrial era and with a probable return to a rural urbanization, what will change in the way we consider the space in which we live that is failing today? This can be seen as a model of "diffuse/weak urbanity", not so far from the concept adopted by the Romans in the *centuriation*, consisting in an economic pattern that is not spatial, as we pointed out. To do so, it is necessary to rethink the palimpsest that concerns us. The *pattern*, intended as a city model, fully represents the combination generated from different functions that has settled in Emilia-Romagna: agriculture, industry, and rurality. The ideological matrix of this structure is present in the territorial culture, as well as the extent of the food culture known all over the world.

Food results from a specific territorial process that needs to be preserved – not the single food product in itself, but within the complete system that has generated it. Food is the product of a specific place; it is above all a cultural product, and for this reason represents within itself all the factors that designed the concept of the landscape into which is inserted.

This is the result of specific interrelations created inside of the same pattern, and similarly it suffers because of the urban planning process. The result is important for the reason that new systems and patterns of territorial urbanization must be preserved as energy resilient producers. This readings linguistic models allow the generation of a series of combinations from regional scale and urban settlement distribution through to the single ornamental element (Alexander, 1977). Being ecologically rational requires recognition of a place as a *socio-natural* construct, calling for the realignment, more specifically, between nature, quality, regional, and local, producers and consumers. Even though many of Emilia-Romagna's food products are available and well-known all over the world, they are generally identified on the basis of a single product, not as a general part of a complex system.

For this reason, the concept of *foodshed* reconstructs the geography of the food systems by compelling social and political decisions on food to be oriented within a specific delineated space (Feagan, 2007).

Physical and psychological distance between food production and consumers creates a tragic disconnect between the general public and the social and environmental consequences of food being grown and eaten. "Food, community and place are complexly intertwined in our lived worlds and across time" (Duruz, 2005).

As we have seen, we cannot admit such a clear separation between rurality and urbanity; the region here represents a perfect example. Reconsidering this gap will help to integrate new development strategies.

The aim of this chapter, outlined by the framework which is based on the region, is to propose and then introduce a new level into urban/sub-urban planning policy, using the food system or foodshed concept as an indicator or key factor. Abandoning the ideological thought of regionalism as the right way of creating synergies, this chapter focuses instead on thinking about food as the representation of the culture of this particular land. In this way, Via Emilia remains a possibility for natural, social, and economic integration.

Bibliography

Alexander, C., 1977, *A pattern language: towns, buildings, constructions*, New York: Oxford University Press.

Alfieri, N. and Schmiedt, G., 1989, *Atlante aerofotografico delle sedi umane in Italia, III – La centuriazione romana*, Florence: IGM.

Allen, S., 2013, "Landscape infrastructure," *Area 127* Identity of the landscape, 18–23.

Arendt, H., 1958, *The Human Condition*, Chicago, IL: University of Chicago Press.

Barham, E., 2003, "Translating terroir: the global challenge of French AOC labelling," *Journal of Rural Studies* 19, 128–138.

Bazzani, A., 2004, *La centuriazione romana nell'agro romagnolo. Le campagne matematiche di Romagna*, Cesena: Società Editrice il Ponte Vecchio.

Bhatia, N., 2012, "Resilient Infrastructure," *Goes Soft, Bracket Series 2*, Barcelona: Actar.

Bhatia, N. 2013, *The Natural Ecology: Soft Infrastructures*. Rice University, Scholarly Interest Report. http://report.rice.edu/sir/faculty.detail?p=fff785e4682e9a00252769676057f5ec

Corboz, A., 1985, "Il territorio come palinsesto," *Casabella* 516, September, 22–29.

Duhl, L. and Powell, J., 1963, *Man and the Environment: The Urban Condition*, New York: Basic Book.

Duruz, J., 2005, "Eating at the borders: culinary journey," *Environment and Planning D: Society and Space* 23: 51–69

Feagan, R., 2007, "The place of food: mapping out the 'local' in local food systems," *Progress in Human Geography*, 23–42.

Forni, G., 1995, "Le colture agrarie padane e la loro produttività all'epoca della romanizzazione," *Rivista Archeologica dell'Antica Provincia e Diocesi di Como* 176: 17–82.

Forman, R.T.T., 1986, "The emergence of landscape ecology," in Forman, R.T.T. and Godron, M. (eds), *Landscape Ecology*, New York: John Wiley & Sons.

Forman, R. T. T., 2005, *Land Mosaics: The Ecology of Landscapes and Regions*, Cambridge: Cambridge University Press.

Gambi, L., 1950, *La casa rurale nella Romagna*, Firenze: Olschki.

Gambi, L., 1961, *Critica ai concetti geografici di paesaggio umano*, Faenza: Fratelli Lega.

Gambi, L., 1989, "Insediamenti e infrastrutture rurali in Emilia Romagna," in Adani, G. (eds), *Insediamenti rurali in Emilia Romagna Marche*, Milan: Consorzio fra le Banche Popolari dell'emilia Romagna e Marche.

George, P., 1956, *La champagne*, Paris: PUF.

Guermandi, M.P. and Tonet, G., 2008, *La cognizione del Paesaggio Scritti di Lucio Gambi sull'Emilia Romagna e dintorni*, Bologna: Bononia University Press.

Guattari, F., 2000, *The Three Ecologies*, London and New Brunswick: Athlone Press.

ISTAT (Istituto Nazionale di Statistica (Italy)). 2014, Quality Agro-Food Products. www.istat.it/en/archive/131530

Marchesini, M. and Marvelli, S., 2010, *Ricostruzione del paesaggio vegetale e antropico nelle aree centuriate dell'Emilia Romagna attraverso le indagini archeobotaniche*, Agri Centuriati: an international journal in landscape archaeology, Pisa-Roma: Fabrizio Serra Editore.

Marsden, T., 2004, "The quest for ecological modernisation: re-spacing rural development and agri-food studies", *Sociologia Ruralis* 44 (2): 129–146

Mazzotti, S., 2003, *Biodiversità in Emilia-Romagna: dalla biodiversità regionale a quella globale*, Ferrara: Museo Civico di Storia Naturale and Emilia Romagna.

McHarg, I. L., 1995, *Design with Nature*, New York: Wiley.

Moussavi, M. and Gareth, D., 2010, *Ecological Urbanism*, Zurich: Lars Muller Publishers.

Potteiger, M., 2013, "Eating places: Food systems, narratives, networks, and spaces," *Landscape Journal* 32(2), 261.

Regione Emilia-Romagna, 2015, *Programma di Sviluppo Rurale dell'Emilia-Romagna 2014– 2020: Italy – Rural Development Programme (Regional)* – Emilia-Romagna. Version 1.4, June 2015. Bologna, Italy, Direzione Generale Agricoltura Economia ittica ed Attività faunistico-venatorie. Regione Emilia-Romagna. http://agricoltura.regione. emilia-romagna.it/psr-2014-2020/doc/testo-del-psr-e-allegati/testo-definitivo-e- completo-del-psr-2014-2020/testo-definitivo-e-completo-del-psr-2014-2020/view

Rowley, T., 1996, *Rural development research: a foundation for policy*, Westport: Greenwood Press.

Schmiedt, G. 1989, *Atlante aerofotografico delle sedi umane in Italia. Parte 3, La centuriazione romana*. Firenze, Italy: Istituto geografico militare.

Sheppard, L., 2013, "From site to territory," *Goes Soft, Bracket Series* 2. Actar: 175–180.

Via Emilia Research Agency, 2014, viAEmilia: Food Cultural Heritage. www.laviaemilia.com

4 Foodscape gastropolis New York City

Arnold van der Valk

Introduction

Men slurp banana pudding from the hips of a black-latex-clad woman lying on a table in a room filled with liquid nitrogen smoke. This is just a single episode in the culinary life of New York City (Parasecoli, 2009). Elsewhere, a long queue of African-Americans is patiently waiting for a soup kitchen in Harlem to open. The temperature is 35°C in the shade. These two images are at the extreme opposites of the spectrum of the subject of food.[1] The role of food in planning sustainable forms of land use in the United States – and in New York City in particular – is the theme of this chapter.

The culinary culture of New York City mirrors the ethnic and social structure of the population of this metropolis. In other words, you are what you eat. New York City is sometimes referred to as a 'gastropolis' (*gastér* is the Ancient Greek word for 'stomach'), a metropolis where life revolves around eating and being eaten (Hauck-Lawson & Deutsch, 2009; Lee & Audant, 2009; de Silva, 2009). This knowledge is essential for planners and designers who want to promote sustainability; the food chain is the key to a more sustainable way of life. Spatial planning can play an important part in this in terms of design, such as establishing frameworks for combining and separating types of space at various levels of scale. In the state of New York, this primarily occurs at the level of the local community (town, borough), which makes it difficult to keep an eye on the bigger picture (city, county, region, state, metropolitan area, watershed, climate zone). This is just a single observation after a three-month visit to New York City and the Hudson Valley.

The investigation described below focuses on a region with a radius of about 100 kilometres from downtown Manhattan. This region includes New York City (with its five boroughs of Manhattan, Queens, Brooklyn, the Bronx, and Staten Island), with a population of more than 8 million, plus another several million people in the New York metropolitan area extending into four or five states. The Hudson Valley, which stretches north from New York City up past Albany, is used as a sample for this metropolitan area. By selecting this river valley, it is possible to look at conurbations, expanses of water, agricultural fields, meadows, and forests. The Hudson River Valley is part of the metropolitan landscape of New York City.

The label 'metropolitan landscape' refers to the fact that land use in this area can only be indicated by constantly referring to the economic strength, lifestyles and patterns of behaviour of the residents of the metropolis in question. The metropolis is an economic, social, and political force field, which shapes and moulds the landscape in a constant process of change. In this context, 'landscape' is seen as a complex of physical characteristics in the environs, which give visual expression to the way people and animals use the land, constructed objects, vegetation, water, and air.

A central question in this chapter relates to the way in which the food system does or does not have an impact on the metropolitan landscape. Changes in patterns of behaviour and land use (which are closely connected to the food chain in New York City and its surroundings) will be examined in detail. Specifically, we will look at whether there are signs of a change or transition into a more sustainable way of dealing with the environment under the influence of changing ideas about the quality of food. The findings are based on conversations with experts, visual observation, websites, and a study of the literature.[2] The motivation for this exploratory study is the idea that the developments in New York City may provide inspiration for planning and designing the metropolitan landscape in the Netherlands and Western Europe (LaBelle, 2005a).

Figure 4.1 Spring 2011: tokens (food stamps) are accepted at the Union Square farmer's market; poor people have limited access to fresh local products due to high prices as compared to processed food on offer in supermarkets

Figure 4.2 Spring 2011: sign indicating the presence of the newly established Urban Farm at Battery Park in the most expensive part of Manhattan

Figure 4.3 Spring 2011: mom educates the kids in a garden in Manhattan's Lower East Side

Food and identity

New Yorkers are proud of the multicultural character of their city. On the City Hall television station, Mayor Michael R. Bloomberg tells viewers on a daily basis that immigrants promote the economic wellbeing of the city. New York City is home to dozens of nationalities. It is a city of ethnic minorities, which maintain their own languages and culinary traditions (Hauck & Deutsch, 2009). At the same time the melting pot is at work, and the results are astounding. A Dutch planner who works for a Business Incentive District in a management position said that two of her employees, one with an Asian appearance and the other one dark-skinned, requested time off during Jewish holidays. Upon questioning, both could prove that they had Jewish mothers. A great many white New Yorkers proudly mention their Dutch ancestry in conversation, though their last names (Luce, Patterson) give no indication of this background. Sometimes they can trace their Dutch roots back 300 years, as in the case of landscape architect Michael van Valkenburgh or lawyer Edie Stone (Brinckerhoff).

Food has an important role for the residents of the city as an expression of the identity of their own ethnic community. Chinatown and Little Italy are the best-known examples, if not the most relevant. Less familiar to tourists, but at least as interesting, is the concentration of Latin American restaurants around Roosevelt Avenue and 37th Street in Queens. The most fascinating phenomenon in this respect is the rise of New York City variations on fusion cuisine. This process has resulted in such foods as chocolate croissants prepared Hungarian style and served in gigantic American portions. Another example is a New York pizza served on an Argentinian folded maize pancake and loaded with Italian sauce, feta cheese and spices. Or Yemeni empanadas, or a dish advertised as 'Italian sushi', which looks more like tapas (de Silva, 2009).

New York City has many faces. For those with jobs paying enough money for them to afford an apartment, life is good. For this group, an estimated 60 per cent of the more than 8 million residents, as well as for most tourists, New York City is a hip, active, green metropolis that is rich in food. This is also the image that outsiders have of the city. For the remaining 40 percent, it is a hungry, unhealthy, and segregated city – but a city, which does offer the possibility to improve your own fate and that of your neighbourhood. It is a socially divided city. The division between rich and poor is evident in the appearance of the neighbourhoods and the lifestyles of the residents. The following sections consider both sides of the city as a background to a discussion regarding food and planning.

The hungry, unhealthy, and segregated city

In 2008 (before the capital market crisis), more than one third of all New Yorkers spent more than 50 percent of their gross family income on rent. At that time, the number of homeless people was estimated at 35,000. More than 1 million people (12.5 percent of the population) is dependent on food stamps and soup kitchens. In the poorest neighbourhoods, found in the boroughs of Brooklyn and the Bronx, that figure rises to more than one half of the population. More than 20 percent of New York City residents earn an income below the official, federally established poverty threshold. In some neighbourhoods, such as in the South Bronx, East Harlem in Manhattan, and Bedford-Stuyvesant in Central Brooklyn, 40 percent of the population live below the poverty threshold (Angotti, 2008). In these parts of the city, according to city government statistics, unemployment is around 25 percent. Because these statistics are used to determine the amount of federal subsidies for food stamps, unemployment projects, etc., they are debated. According to estimates by social workers, the actual level of unemployment is more than 50 percent. The official unemployment figure for New York City, according to the state of New York, is 8.7 percent of the labour force. It just depends on how you measure it!

Poverty is concentrated in the neighbourhoods dominated by African-Americans and people with a Latin American background. The average income of residents of African-American neighbourhoods is 55 percent of the average income of residents of predominantly white neighbourhoods (Angotti, 2008). Segregation along racial and ethnic lines coincides with segregations along income lines. Race and ethnic background always have an effect on everyday life, even if they are rarely spoken of openly. If you are African-American and you live near the Crown Heights/Utica Avenue subway station in Brooklyn, there is a 50 percent chance that you are unemployed. In that case it is likely that too much of your income is spent on rent and that you have too little money to spend on healthy food (if it's even possible to get fresh products anywhere nearby). There is a large chance that you regularly stand in line at the soup kitchen run by the local church. And finally, you are at risk of obesity, diabetes, asthma, or other diseases related to poor nutrition. More than 3.1 million New Yorkers are overweight and no less than 20 percent of the population suffers from obesity (!).

In some neighbourhoods of the South Bronx, it's nearly impossible to find a fresh head of lettuce or an orange. Paradoxically, this neighbourhood borders on the largest fresh food market in the world, the Hunts Point Market. The Hunts Point Market is a wholesale food market, which is not accessible to private buyers. These neighbourhoods, home to 37 percent of the population of New York, are known as 'food deserts' (Amstrup *et al.*, 2011).

The problems don't stop with chronic poverty, poor housing, unemployment, and low-quality food. Highways built by the urban planner Robert Moses in the 1950s and 1960s run through the poor neighbourhoods of the Bronx, Harlem, and Brooklyn. These highways create huge amounts of air pollution. On top of this, 30 percent of all the city's waste and 70 percent of all its sewage is processed in the poor neighbourhoods.

In this desolate environment, urban agriculture has an important role. There are hundreds of community gardens. Some of these specifically advertise themselves as urban farms. These gardens are usually located on municipal land and are tended by a group of local residents for whom they represent both a means of sustenance and a source of relaxation. These gardens originated in the 1970s, at a time when New York was on the verge of bankruptcy. Municipal amenities had been reduced to an absolute minimum. The effects of that were that the police and firefighters couldn't respond to calls as often, and entire neighbourhoods came under the control of drugs gangs. Owners of apartment buildings saw their rental incomes decrease to the point that maintenance costs and city property taxes exceeded the incomes. As a result, many property owners deliberately burned their buildings down in the hope of receiving payouts on their insurance. This land then became the property of the city, which then had a claim against the prior owner for taxes owed.

Those remaining in the neighbourhood tried to make the best of a bad situation. They set up neighbourhood coalitions and started cultivating the plants that grew naturally among the rubble of the collapsed buildings. Artists calling themselves 'green guerrillas' threw seed bombs over the fences around desolate ruins in the hope of encouraging plant growth (Tankersley, 2009). This made these spots less attractive for dumping garbage (municipal garbage collection wasn't functioning efficiently) or as hangouts for drug addicts. Over time, between 1970 and 1990, the rubble was cleared away, and many gardens were turned into urban oases.

Primarily white neighbourhoods underwent a process of gentrification, in which well-to-do, well-educated whites moved in. This caused rents to rise and made it attractive for project developers to build on empty lots. This happened on a large scale in Manhattan's East Village and in that part of the Upper West Side, which borders Harlem.[3] Around 1990, the mayor at that time, Rudolph Giuliani, attempted to auction off all of the community gardens, which had arisen spontaneously and sell them to project developers. This brought about a revolt among the community gardeners, who, with their family and friends, numbered about 100,000 voters.

In 2002, Giuliani's successor, Michael Bloomberg, decided that it would be unwise to pursue the conflict any further. On behalf of the city, he turned most of the plots containing community gardens over to land trusts. The transfer contract contained a clause that the land be conditionally reserved as a garden for a certain period (specifically, Bloomberg's term of office). However, this reservation is not established in the city's zoning ordinances. The same is true for the parks. The preservation of the community gardens is not entirely altruistic on the part of the city and developers, because research has shown that community gardens increase the value of surrounding property, and therefore yields from city property taxes (Voicu & Been, 2008). The gardens in the areas, which have undergone gentrification are little paradises, where members of the local garden group invest huge amounts in seeds, wrought-iron fences, and works of art.

In the poorest neighbourhoods, the community gardens are no less important, but they are usually very different in appearance. Here, most of the ground – or the pots containing humus and compost – is used to grow vegetables and fruit. Recently, community gardeners have also been permitted to keep chickens and bees. Many gardens in the Bronx and Brooklyn are referred to by their users as 'urban farms'. This is in the hopes of being listed as an agricultural business in the agricultural census. If a garden shows a turnover of at least US$1000 annually, it falls under the category of 'agricultural business'. The owner and/or leaseholder (in this case the gardening group) is then eligible for tax deductions and even for farm subsidies.

Many of these community gardens in poor neighbourhoods have a manager, often white, who is paid by the New York Botanical Garden. The gardeners are required to attend free classes in subjects such as composting, using basic farm tools, soil science, aquaponics, and setting up community supported agriculture (CSA). Some initiatives, such as the gardens run by the organisation Added Value in African-American neighbourhoods like Red Hook in Brooklyn, have developed into multifunctional businesses utilising advanced techniques.[4]

The Added Value garden in Red Hook consists of compost beds laid on an asphalt surface (a former basketball court). The gardeners work with compost (plant waste from parks and streets) on a large scale, delivered by the City Department of Sanitation. There are plans to install a closed water system, in which wastewater will be used for farming fish (tilapia). Those working at the Added Value farm include white social workers, a few paid employees (both white and African-American), and African-American and white volunteers, including a group of people wearing brown uniforms. In the summer, the garden hosts a farmer's market on the site, where restaurateurs and people from the nearby tenement blocks can purchase fresh produce. The farm also has a CSA, which is of great importance to the neighbourhood. The farm's regular staff is dedicated to the educational side of the work in the local community. The farm allows children to learn what fresh food is, how it is grown, and how it is processed. There are lessons in cooking with fresh products. Young people from the neighbouring apartment buildings can receive an education in urban agriculture. Added Value receives income from various sources and has access to city government facilities. Sales of farm produce to private individuals and local restaurants can cover an increasing portion of the costs. The organisation is proud of its sustainable image, and it is visited by hordes of foreign green tourists (including the author of this piece). The management works hard to continue to develop the educational and tourism aspects in a commercially responsible way. In order to achieve that goal, they have opened a vegetable garden and composting centre on Governors Island in Upper New York Bay. In the future, the farm may well form part of an ecological education centre for an international audience. In order to realise that project, the organisation called in the assistance of Adriaan Geuze and his urban design bureau West 8, and Will Allen, an African-American public hero from Milwaukee.[5] Will Allen started a successful urban farming business, an education office, and a national training centre under the name 'Growing Power'. In 2008 he won the McArthur Fellowship for his pioneering work in urban farming. This is a prestigious award involving a grant of $500,000.

The hip, active, green metropolis, rich in food

'After life in New York, everything else is so boring', is a common expression uttered by foreigners who have lived in New York City for any length of time. Manhattan's theatres and museums, the concerts in Central Park, the infinite

Figure 4.4 Spring 2011: projects (social housing) in the Bronx where a substantial part of the black and Latino population lives in poverty and community gardens make produce accessible for low-income households in food deserts

Figure 4.5 Spring 2011: would-be workers on one of Ben Flanner's roof-top farms receive instructions on a Saturday morning; labourers pay a $50 fee to join the crop mob

variety of restaurants, the views of Manhattan, the street musicians – these are just some of the things making life in New York, and the cost of it, worthwhile. Residents and tourists with time and money jog or bicycle in Central Park in their vivid designer clothes, drink a Belgian(!) beer on a café terrace in the Meatpacking District, or spend their free Saturday in a 'crop mob'. 'Crop mobbing', participating in a crew on an agricultural business on the roof of an office building in Queens or Brooklyn with a breathtaking view of Manhattan, is all the rage right now among students. It costs around $45 to participate – in other words, you are not paid to work; you pay to be allowed to work.[6] This is one facet of the recent fascination, which young, progressive New Yorkers have for the quality of food and the environment. Food is one of the most popular themes on local television, in conversations on café terraces, and in any number of columns in *The New York Times*. There is interest in every aspect of the food chain, from land use, design, transport, and processing to distribution of food, and waste processing. The central theme is promoting relationships between producer and consumer and making the processes behind food more visible.

This is the context within which the shift to forms of more sustainable agriculture in and around New York City is taking place. At the risk of creating a caricature of the situation, I wish to postulate that the new agriculture is taking place in light of a shift in thinking among a rapidly growing group of consumers in New York City. These consumers are concerned about the quality of the food and its consequences for the environment. On a large scale, these consumers purchase their food at the expensive Whole Foods Market chain or at the local farmers' market. The number of farmers' markets has risen from a few dozen in 2000 to more than 120 in 2011. Sustainable agriculture is just one aspect of the broad social movement for a more sustainable and healthier lifestyle, which can be compared to a historical movement that occurred around the turn of the century in favour of social reform (better and more affordable residences, parks, sports fields, school gardens). The current movement is still in its infancy, but in New York City it has gained the attention of politicians from both the left and the right sides of the political spectrum. The current mayor, Bloomberg, and two potential mayoral candidates for the 2013 elections, each published policy plans and reports on a more sustainable food chain in New York City in 2030 (Bloomberg, 2011). Michelle Obama's vegetable garden at the White House in Washington fulfils an important symbolic function, and it has inspired the establishment of urban farms in places like Battery Park at the tip of Manhattan, on the grounds of New York City Hall near Wall Street, and on the roof of the car park belonging to the administrative building of the New York City Department of Parks and Recreation on Randall's Island.

The ins and outs of sustainable agriculture in the city are explained by a young farmwoman on the roof of an urban farm in Queens, who has studied communication science and agricultural economy, specialising in tax law. Urban agriculture is sustainable, hip, young, socially relevant, and entrepreneurial. The new generation of urban agricultural entrepreneurs has a design for vertical farms on the drawing board, and right now it is being tested in Milwaukee. The ecologist Dr

Dickson Despommier of Colombia University is delighting young architects with futuristic concepts for agricultural apartment buildings in Manhattan (Despommier, 2009). The question posed in this by critical scientists is whether these developments will also benefit employment opportunities and nutrition for people in poor neighbourhoods and for small farmers in the surroundings of New York City. In all honesty, from the top floor of the Rockefeller Building, known as the 'Top of the Rock', even with binoculars it's not possible to see any signs of a transition in which 'roof-top farms' and 'vertical farms' play a part. Upon investigation, green and blue spots on a rather low roof on 50th Street in Manhattan turn out to be a collection of flowerpots surrounding a swimming pool.

New York City's food system

The current food system is anything but sustainable. There are issues such as risks to the continuity of the supply (safety), a lack of availability of healthy and affordable food in the city's poorer neighbourhoods, an excess of unhealthy foods, and products that have been transported there from other parts of the world. For most people, food is an abstract product. There is an increasing group of people, which include the current president and his wife, who see poor nutrition as a threat to public health. The city's potential resources (energy, nutrients for plants and animals) are unused. The metropolis' rubbish ends up at dumps spread over several states. The organic portion of that rubbish is subject to digestive processes, which take place in an oxygen-poor environment, thereby contributing to the greenhouse gases in the atmosphere. In short, it is a miserable situation, which in the opinion of New York City politicians must be changed (Bloomberg, 2011).

Most of New York City's food supply travels through a gigantic wholesale market in the Bronx and via mega-warehouses belonging to a few chain stores (Cohen & Obadia, 2011). The transport of raw materials for agriculture and processing of end products are in the hands of a few agricultural mega-companies with branches around the world, like Monsanto, Kraft, DuPont, and Unilever. These raw materials and products are transported via the already overburdened highway system, a few gigantic terminals in the New Jersey Harbour, and three airports. An attack on a single distribution centre, bridge or airport could totally disrupt the metropolis's food supply; this was demonstrated in a strategic study carried out by the RAND Corporation and the simulation exercise Silent Prairie, a war game developed by the Department of Homeland Security (LaBelle, 2005b).

Food in conventional supermarkets is inexpensive – often less than half the price of agricultural products at the Union Square farmers' market in the heart of Manhattan. The agricultural lobby in Washington uses the stream of cheap food products as a strong argument for maintaining, and even expanding, the complex system of tax deductions and agricultural subsidies. The conventional industrial system of production, trade, processing, and distribution is a cornerstone of the American agricultural policy, and it is periodically reinforced in the Farm Bill, a key agricultural law. Even in this period of massive government deficits, the American government has not managed to reduce agricultural subsidies. The policy is largely dictated by a handful of agricultural mega-companies and regional farmers' organisations. Their contributions to the election campaign funds of politicians in every party mean that they maintain a firm hold on political representatives, particularly on those from the Mid-Western agricultural states.[7]

Maintaining the status quo in agricultural politics comes at a high price, in the form of deteriorating soil, water, and air quality. Champions of healthy food consistently start their exhortations with a reference to the average distance travelled by a tomato or a piece of meat in North America from the moment of harvest to the supermarket cooler. This distance is represented by the magic number 1,500 miles (McWilliams, 2009). The implication is that transporting food by air and over land makes a substantial contribution to the problem of greenhouse gases and global warming. The biodiversity and beauty of the landscape are disappearing rapidly. In large parts of the United States, the infrastructure to support agriculture on a smaller scale and more sustainable agriculture has been destroyed (Glynwood, 2010). In many developing countries, agriculture is suffering severely as a result of the dumping of agricultural bulk products from America and Europe. The most important raw material for conventional agriculture, oil, is becoming scarce and therefore expensive.

Too often, food is harmful to people's health. Within just a few years, obesity and malnutrition have grown into political problems of the first order. Treating its medical consequences costs the American taxpayer $100 million per year. It's an illness of the poor, and is connected to the proliferation of fast food. Excessive use of antibiotics in livestock farming forms another threat to public health – as was evidenced by the recent *Escherischia coli* outbreak in Germany. The intensive agriculture sector is a laboratory for resistant bacteria, which irrevocably spread throughout the environment despite environmental regulations.

The dominant conventional food system, with its dependence on global trade, industrial production, agricultural subsidies, and monopolistic tendencies, offers few opportunities for growth and recovery in the form of small-scale, regional, and sustainable agriculture in New York City or elsewhere in the world. Many farms are forced to produce more intensively and on a larger scale because the companies providing the seed, manure, and capital are paying less and less for products. The banks, which farmers have to approach for capital at the beginning of the season, hold a knife to the throats of the small and medium-sized farmers. Many farmers are faced with the choice of 'growing big or getting out'.[8] In the long term, the continued existence of entire agricultural communities is at risk because the quality of life in villages is increasingly under threat from pollution, health risks, and the disappearance of amenities. In the Hudson Valley, New York City's back garden, all these processes and mechanisms are active. The following section will show, however, that rather than being nothing but a vale of tears, a few bright spots are arising in agriculture around the city. This raises the question of whether a transition in agriculture is on the horizon. If this is so, what are the possible consequences for the landscape?

Agriculture and landscape in the Hudson Valley

New York is blessed with an environment (climate, soil, air, water) offering favourable circumstances for agriculture. As early as the seventeenth century, the Dutch West India Company seized the opportunity to cultivate agricultural products in its colony New Netherland (modern-day New York, New Jersey, Connecticut, and Delaware) so as to provision Dutch ships. In the first decades of the twentieth century, more than half of all food consumed in New York City had been cultivated within or just outside the municipal boundaries. Although there are no exact figures kept for current food production in and near New York City, it is not unreasonable to assume that that number now amounts to no more than a few percent.[9]

Until 2005, it appeared that the future of food production and the continuation of farms in and around New York City were doomed. The food processing industry, by contrast, still has an important role within the business world. The number of people completely supported by a farm in New York City or the Hudson Valley has been falling for decades (Glynwood, 2010). The majority of the farmers remaining are approaching the retirement age. A large portion of their incomes comes from sources other than agriculture, such as an office job, camper storage, corn mazes, and farm golfing. This is worrisome to organisations that work for the environment, water collection, nature, and landscape. The geographical situation – the location in the metropolitan landscape of New York City – is both a blessing and a curse. The curse has to do with the high land prices because of the relative scarcity of farmland. In 2007, 17 percent of the land in the Hudson Valley was used for agriculture, as opposed to 24 percent for the entire state of New York. This primarily has to do with the relatively large and increasing surface areas of parks, nature reserves, water-collection areas, recreational areas, and urban expansion. The blessing of the proximity to New York City goes hand in hand with the explosively growing demand for high-quality food. Demand is growing, but unfortunately the remaining farmers are presently unable to meet it. This is mostly the result of deteriorating infrastructure in the past decades of scale increases. This deteriorating infrastructure includes veterinarians, suppliers of animal feed and fuel, repair shops for agricultural equipment, and food processing companies such as slaughter-houses.[10]

New York City's political elite and the readers of *The New York Times* see the decline of the agricultural sector as a major problem. The Hudson Valley is New York City's 'back garden', where a great many people take trips, visit restaurants in the summer, or even have a country home. The concerns of the urban elite, as articulated in the reports by the farmers' interest group Glynwood, particularly relate to the risks to supplying the metropolis with healthy, local, affordable food. The problems faced by the predominantly small, sustainably producing farmers, who also have an important role as stewards of a very diverse and intimate landscape, are evidenced by the shockingly high prices for healthy food (Glynwood, 2010). In a typical example from the life of the author of this piece two, admittedly beautiful, organic heirloom tomatoes each weighing nearly two ounces, cost $9 at the Union Square farmers' market. A 200-gram piece of farmhouse cheese cost $11.[11]

The data from the last agricultural census, in 2007, and data collected by Glynwood indicate a shift. Certain striking developments may indicate the beginnings of a transition. The figures are not self-evident, so I will briefly explain them and place both the negative and positive aspects in context.

The number of farms in the Hudson Valley decreased by 3 percent between 2002 and 2007. The surface area of land used for agriculture decreased by 10 percent. This includes land, which lies partially fallow or is overtaken by forest. Farms in the 'large businesses' category (500 acres or more) have seen their land decrease the most. Dairy farms are facing the severest decreases. In the dairy sector as a whole, the number of farms has decreased by 27 percent as a result of sharply falling prices over the last few years. At the same time, production costs have increased by 21 percent. This has to do in part with high land prices and the difficulty of buying or renting extra land. Farmers in the Hudson Valley struggle with very small margins in a cost-yield ratio of 94 to 100. The profit margins in the predominantly small farms are very small and the farmers' incomes are predominantly low. The small profits can easily change into losses. Fewer and fewer farmers list farming as their primary profession. Most farmers depend on income earned outside the farm. Nearly half of all farmers indicated that the farm is related to post-retirement activities or has the character of a supplementary source of income or a hobby.

According to the 2007 agricultural census, there are still 5,326 farmers left in the Hudson Valley. Because of their relationship with the landscape, it is worth examining this group and their future in more detail. First of all, this figure must be put into perspective with reference to the broad definition of an agricultural business. The United States Department of Agriculture defines it as businesses with a minimum annual turnover of $1,000, as well as businesses with the potential to produce that amount. This definition includes many hobby farmers and community gardeners. On the basis of this figure as well as the results of a survey, which it carried out under farmers, Glynwood estimates that there are 4,000 farmers in the area. The members of this group have a farm that is at least as large as a football field and that generates a substantial part of the family income. These farmers have an important role in maintaining the intimate landscape with its forests, wildflower meadows, orchards, ponds and wooded banks. In other words, there is no shortage of problems.

The positive aspects relate first and foremost to the growing demand for healthy and locally produced fresh food in the metropolis. Because of the growing desire of consumers to have contact with food producers ('know the face of the farmer'), the regional value of direct sales from farmer to consumer in the Hudson Valley has risen sharply – from $11 million in 1997 to $25 million in 2007. In 2010, 17 percent of the farms in the region received all or part of their incomes from direct sales through CSA or farmers' markets. The demand for healthy fresh food is growing in particular among those with the lowest incomes, who receive food stamps from the government in place of welfare payments. Purchases made with food stamps at organic shops and stands make up the most rapidly growing segment of the market for organic food in the United States. Sales of brand-name organic food increased from 15 percent in 2007 to 18 percent in 2010. Garden centres are reporting massive growth in organic products in recent years, and this trend has not really been affected by the economic crisis.[12] These trends can also be seen in New York City. However, it must be remembered that in absolute terms, the movement to consume healthy food is still a marginal phenomenon in the United States as a whole.[13] New York City and other metropolises such as Chicago, Seattle, and San Francisco are at the forefront of this.

In addition, the number of farmers, in particular women and young farmers, increased slightly between 2002 and 2007 (even discounting the increase which is a consequence of the new definition).[14] There is also a substantial increase in direct sales of products from the farm. The amount sold from farms rose by 36 percent between 2002 and 2007. Farming-related income increased by 62 percent, with the largest growth occurring in income from agro-tourism.

Market research shows that farmers in the metropolitan landscape of New York City can take advantage of a wealthy and growing market for quality food (Glynwood, 2010). The consumer population is relatively wealthier (and more ethnically diverse) than in other American metropolises such as California's San Francisco Bay Area and Southeastern Pennsylvania. Despite these positive perspectives, the turnover and profits are relatively low in the Hudson Valley. The essential question is why the potential arising from the increasing demand is not being utilised to its full extent.

According to research carried out by Glynwood, the biggest obstacles are the high costs of operational management; the lack of availability of land for expanding farms; and the lack of infrastructure for the production, processing, and sales of regional agricultural products. Production is not sufficiently integrated with the other elements of the chain: product processing, rental and repair of agricultural equipment, distribution, and marketing. The lack of infrastructure causes operational management costs to rise. Equipment must be obtained from far away. There are no slaughterhouses in the region and there are few places where the products can be processed and packaged.

However, at the time of writing, innovative projects to strengthen the infrastructure have been initiated. A few examples follow, offering an idea of the direction in which agriculture in this region is headed.

The first innovation deserving of attention is called Pampered Cow. This is a new business which helps small-scale cheese makers efficiently market their products. Right now, each agricultural entrepreneur uses his or her own transportation to get to farmers' markets several days a week. This costs both time and money. Pampered Cow utilises the principle 'build relationships, not bricks-and-mortar'. The business keeps track of which refrigerated trucks belonging to large companies have extra space in their journeys to and from the markets. This space is purchased on behalf of the farmers for transporting small shipments of cheese. If desired, the cheese is sold as a consignment. The business mediates between producers and buyers and makes sure that the farmer is paid on time.

Another project is called Farm to Table Co-Packers. This company addresses an issue faced by many small producers – keeping perishable products fresh longer – by cooking them, freezing them, salting them, pickling them, or otherwise extending their shelf life. Farm to Table remodelled the main kitchen of an abandoned IBM building into a processing centre for agricultural products. This enables small business owners to add value to their products. The company actively looks for buyers of products such as tomato juice, pickles, and frozen vegetables. Buyers can be found in school lunchrooms, the restaurants of municipal organisations, and nursing home kitchens.

Hudson Valley Fresh is a company that processes fresh milk into dairy products. The participating dairy companies commit to maintaining higher quality standards than the regular dairy processing companies. The business has a cooperative character and responds to the demand from New York City for high-quality food with a regional brand.

One final initiative, which deserves a special mention, is Glynwood's design for and use of a mobile livestock slaughterhouse. Contrary to the expectations of sceptics, this unique initiative was recently given approval by the United States Department of Agriculture.

What part can planners and designers play?

On the surface, spatial planning only has a marginal role in the process of changing the agriculture of New York City's metropolitan landscape. In the end, will the market determine the future of agricultural land use? Yes, because ultimately the economic forces of supply and demand are the deciding factors. From the point of view of sustainability, the changes in those forces appear rather favourable at the moment. In this process, planning and design have a facilitating role, particularly at the municipal level, where the zoning plans are determined (Cullingworth and Caves, 2003; Delaware Valley Regional Planning Commission, 2010).

The importance of a well-functioning food system in the context of 'smart growth' and 'comprehensive planning' in sustainable cities was recognised a few years ago by the American Planning Association (APA), the professional organisation of American planners and urban developers. The APA has published a policy guide for local and regional food planning (American Planning Association, 2007). This document contains an in-depth analysis of the importance of the spatial component of the food system in the context of land use in cities and rural areas. Among the topics it covers are maintaining biodiversity; limiting air, water, and soil pollution; aiming for sustainable economic development; and strengthening agricultural communities (tax bases, an awareness of the importance of agriculture). Landscape is mentioned in passing. The APA makes a number of concrete recommendations, all of which are utilised in New York City. It is necessary that planners take stock of the complex process of the food chain and the spatial effects connected to it. Politicians and interest groups could then use this data to back up their demands and needs, particularly demands for financial resources and regulations. This implies a call to professionals to work together with urban and rural residents to chart the food stocks, the potential for food production, and the risks in terms of the fresh food supply.[15]

Taking stock of local and regional food systems can make an important contribution to the local economy and help preserve and create jobs.[16] This can help keep farmers from leaving preponderantly agricultural rural areas. Examining the potential in local and regional food systems can make a significant contribution to improvements to the American health situation. Possibilities include measures in the context of comprehensive plans and zoning by local communities. For these reasons, the APA supports planning and designing food systems, to the extent that this fits within the trend towards ecological sustainability. Drawing up zoning ordinances and environmental effect reports are given as the most important measures. In dealing with food systems, it is of express importance that all stages of the food chain are taken into account, including the processing of agricultural products, transport, and particularly waste management. One final aspect concerns issues of social justice: consumers' access to healthy food. This primarily relates to the availability of fresh fruit and vegetables in the poorer neighbourhoods and towns. One important element involves incorporating attention to food in the education of planners and designers. The APA will include this theme in the requirements to which educational programmes are subjected in order to receive accreditation from the professional organisation.

It is surprising that the professional organisation for planners and designers takes an explicit position in the debate regarding agricultural policy in the context of the periodic revision of the Farm Bill. The policy guide explicitly criticises the agricultural politics of the federal government where this is still dominated by increases in scale. The APA advocates measures at local and regional levels, which can serve as a counterweight to the increases of the scale of industrial agriculture.

This exploratory study into the developments in urban agriculture and the agriculture within the sphere of influence of the gastropolis that is New York City shows that there is still a world of progress to be made on the path to a sustainable food system. Of course the current system is accompanied by social injustice, risks in terms of supplying the metropolis, unhealthy food, decadence, and waste. But there are a great many positive signals counteracting this – signals which could be interpreted as a transition to a new food system in which healthy food, an economically healthy food sector, and sustainable production, processing, transport, and waste processing are the norm. In this system, the urban landscape of Manhattan will be enriched with vertical farms, fresh vegetables will be produced on the roofs, and dog parks will be turned into gardens. The phenomenon of CSA enables young entrepreneurs to produce food in back gardens, public parks, and car parks. Many restaurants are already serving meat from free-range chickens and bok choy grown on a Queens acre farm. For those with a sweet tooth, there's a wide selection of honey from Manhattan roof gardens. Children are involved in food production and caring for the quality of the soil, water, and air from an early age.[17] The traditional agricultural landscape with its small fields, woods, ponds, and brooks is being given a new impulse as a result of the growing demand for healthy locally and regionally grown food. Chicken now comes from Brooklyn, honey from Queens, sheep's cheese from Ulster County, and free-range meat from Orange County. It's still on a small scale, but the seed of sustainable agriculture has been planted. A mechanism has been set into action, carried by women, children, young entrepreneurs, students, and immigrants. Growing tomatoes in pots in front of your house is stylish. Composting organic waste is all the rage. There's something brewing in New York City. If the goal of getting a city of 10 million people to completely change their behaviour is successful, it will have enormous consequences for the global transition to a more sustainable society (Cohen & Obadia, 2011).

One of the lessons from the studies carried out by Glynwood is that the examination of planning activities related to sustainable organisation of the food chain and landscape development should take place from a regional perspective. This is contrary to the American tendency to see planning as zoning land use in and by local communities.

Notes

1 This was observed by the author while cycling through Harlem and the Bronx on 19 June 2011. The queue was observed on East 120th St. at a food pantry in a church.

2 The author spent a three-month sabbatical in New York City at the Tishman Center for Environmental Studies of the New School for General Studies. His host and direct colleague was Dr Nevin Cohen, a specialist in the field of food, environmental issues, and spatial planning in New York City. A complete list of individuals interviewed and books, articles, and websites consulted can be found in the appendix of the report produced by the author. Copies of the statements have also been included.

3 This is the area where the author of this piece lived for three months. The address is 133 Manhattan Avenue, NYC, NY 10011.

4 See: www.added-value.org/ Accessed on 28 June 2011.

5 See: www.growingpower.org/ Accessed on 28 June 2011.

6 For work on the rooftop farms see: www.brooklyngrangefarm.com/ www.takingrootus.com/ Accessed on 28 June 2011.

7 This issue was discussed in *The New York Times* of 23 June 2011, page A14, under the title 'In Battle Over Subsidies, Some Farmers Say No'.

8 Interview with Judith LaBelle on 23 May 2011.

9 Farming Concrete NYC is a project started by Mara Gittleman. She distributes forms among the community gardeners and urban farmers on which they indicate the yields from their gardens (type of product, date, and amount of harvest). The results are statistically calculated and published on a website. The data are then used as input for the Five-Borough Farm Project, which charts the current state of urban agriculture for particular initiatives and municipal policy. See site: http://farmingconcrete.org/ Accessed on 15 June 2011.

10 Interview with Judith LaBelle on 23 May 2010. See also: www.glynwood.org/ programs/modular-harvest-system/

11 www.wholefoodsmarket.com/

12 www.organicconsumers.org/articles/article_22193.cfm

13 The number of organic farmers in the US is increasing. In the first 2008 survey of organic agriculture in the US, 14,540 farms were classified as organic, with a total productive area of 4.8 million acres (as opposed to 2.9 million acres in 2006). The number of certified farms increased from 9,501 in 2006 to 12,941 in 2008 (www.ers.usda.gov/StateFacts/US.htm). In an absolute sense, the surface area for

organic agriculture in the US takes up no more than 0.5 percent of the total area of agricultural land, as compared to 2.5 percent in the Netherlands. In 2005, there were 427 organic farms registered in the state of New York, with a total of 68,864 acres of land. We can compare this to figures for the Netherlands. In 2004, the area used by organic agriculture in the Netherlands rose to 48,155 hectares. The growth up to 2004 was primarily caused by registrations of grassland. In 2004 the area used by organic agriculture slowly shrank to 47,019 hectares by late 2007. In 2008 it grew by 7.3 percent to 50,435 hectares. Most of the expansion took place in the provinces of Drenthe and North Holland. In the province of Flevoland, the area shrank slightly. The total area of agricultural land in the Netherlands was 1.9 million hectares in 2008 (www.agriholland.nl/dossiers/bioland/home.html#omvang).

14 See the documentary *Greenhorns*: www.youtube.com/watch?v=zH7o3fxw6oE Accessed on 28 June 2011.

15 In 2009, the National Association of Development Organisations published the report *Regional Foods Systems Infrastructure* in order to create a regional development organisation. This organisation will serve farmers, food companies, and food transporters who want to work regionally. The report presents three regions in different parts of the US as examples: the Sacramento Area Council of Governments, Southwest Wisconsin, and the Western North Carolina Regional Livestock Center. According to Judith LaBelle of Glynwood, many of the findings perfectly correspond to the Hudson Valley. See: www.ruraltransportation.org/uploads/regionalfood.pdf

16 The lack of data on the yields and the consequences for the local community is a disadvantage. 'Unknown, unloved' – politicians want to see figures.

17 See the website: http://thegreenest.net/2011/01/2010-nyc-urban-agriculture-roundup/

References

American Planning Association (2007). *Policy Guide on Community and Regional Food Planning*. Download: www.planning.org/policy/guides/pdf/foodplanning.pdf

Angotti, T. (2008). *New York for Sale: Community Planning Confronts Global Real Estate*. MIT Press: Cambridge, MA.

Amstrup, I., Boga, M. and Bouyer, C. (2011). *La Finca Del Sur: Uprooting Food Insecurity In The South Bronx*. The New School: New York.

Bloomberg, M.R. (2011). Update to *PlaNYC* 2030: A Greener Greater New York. Download: www.nyc.gov/html/planyc2030/html/theplan/the-plan.shtml

Cohen, N. and Obadia, J. (2011). 'Greening the Urban Food Supply in New York', in: Slavin, M. ed. *Sustainability in America's Cities: Creating the Green Metropolis*. Island Press: Washington DC, pp. 205–229.

Cullingworth, B. and Caves, R.W. (2003). *Planning in the USA: Policies, Issues and Processes*. 2nd ed. Routledge: Abingdon.

De la Salle, J. and Holland, M. (eds) (2010). *Agricultural Urbanism: Handbook for Building Sustainable Food and Agriculture Systems in the 21st Century Cities*. Green Frigate Books: Winnipeg.

Delaware Valley Regional Planning Commission (2010). Food System Planning. Municipal Implementation Tool #18. Download: www.ruaf.org/ruaf_bieb/upload/3304.pdf

Department of City Planning of the City of New York (2011). *Zoning Handbook*, 2011 Edition. The City of New York: New York City.

De Silva, C. (2009). 'Fusion City: From Mt. Olympus Bagels to Puerto Rican Lasagna and Beyond', in: Hauck-Lawson, Annie and Deutsch, Jonathan, eds, *Gastropolis: Food & New York City*. Columbia University Press: New York, pp. 1–14.

Despommier, D. (2009). *The Vertical Farm: Feeding the World in the 21st Century.* Thomas Dunne Books: New York.

Glynwood, (2010). *The State of Agriculture In The Hudson Valley.* Glynwood: Cold Spring, NY. Download: www.glynwood.org/files/2011/02/State_of_Ag_2010.pdf

Hauck-Lawson, A. and Deutsch, J. (eds) (2009). *Gastropolis: Food & New York City.* Columbia University Press: New York.

LaBelle, J. (2005a). 'A Time Of Great Challenge And Opportunity', *Gleanings,* Summer, pp. 1–5. Glynwood Center: Cold Spring, NY.

LaBelle, J., (2005b). 'New Perspectives on Innovation from The Netherlands', *Gleanings,* Winter, pp. 1–5. Glynwood Center: Cold Spring, NY.

Lawson, L.J. (2005). *A Century of Community Gardening in America.* University of California Press: Berkeley.

Lee Perez, R. and Audant, B. (2009). 'Livin' La Vida Sabrosa: Savoring Latino New York', in: Hauck-Lawson, A. and Deutsch, J., eds,. *Gastropolis: Food & New York City.* Columbia University Press: New York, pp. 209–229.

McWilliams, J.E. (2009). *Just Food: Where Locavores Get It Wrong and How We Can Truly Eat Responsibly.* Back Bay Books: New York.

Metcalf, S.S. and Widener, M.J. (2011). 'Growing Buffalo's capacity for local food: A systems framework for sustainable agriculture', *Applied Geography,* doi:10.1016/j.apgeog.2011.01.008.

Nairn, M. and Vitiello, D. (2009). 'Lush Lots: Everyday Urban Agriculture: From Community Gardening to Community Food Security', *Harvard Design Magazine,* no. 31 (fall/winter 2009/10), pp. 1–8.

National Association of Development Organizations Research Foundation (2010). Regional Food Systems Infrastructure. NADO: Washington DC. Download: www.ruraltransportation.org/uploads/regionalfood.pdf

Parasecoli, F. (2009). 'The Chefs, the Entrepreneurs, and Their Patrons: The Avant-Garde Food Scene in New York City', in: Hauck-Lawson, A. and Deutsch, J., eds, *Gastropolis: Food & New York City.* Columbia University Press: New York, pp. 116–130.

Poppendieck, J. and Dwyer, J.C. (2009). 'Hungry City', in: Hauck-Lawson, A. and Deutsch, J., eds, *Gastropolis: Food & New York City.* Columbia University Press: New York, pp. 308–326.

Quinn, C.C. (2010) *Food Works: A Vision to Improve NYC's Food System.* NYC City Council. Download: http://council.nyc.gov/html/action_center/food.shtml

Ruhf, K. and Clancy, K. (2010). *It Takes A Region: Exploring A Regional Food Systems Approach.* Northeast Sustainable Agriculture Working Group: Washington DC. Download: www.ittakesaregion.org/uploads/2/7/7/0/2770360/regional_food_system_working_paper_final.pdf

Srisethnil, A., San Buonaventura, A. and Perry, A. (2011). *Urban Agriculture for Healthy Cities: A Study of Food Access in Brooklyn's Community District 9 and The Programs of BK Farmyards.* The New School: New York.

Stringer, S.M. (2010). *FoodNYC: A Blueprint for a Sustainable Food System.* Office of the Manhattan Borough President: New York City. Download: www.mbpo.org/uploads/policy_reports/mbp/FoodNYC.pdf

Tankersley, G. (2009). *Community Gardens of the East Village: An Oral History Guide to Community Gardens in New York City's East Village.* Grace Tankersley: New York.

Voicu, I. and Been, V. (2008). 'The Effect of Community Gardens on Neighboring Property Values', *Real Estate Economics,* vol. 36 no. 2, pp. 241–283.

Interviews

Interview with Jack Linn, former Deputy Park Commissioner of the NYC Department of Parks and Recreations, on 19 April 2011 at Veselka, 2nd Avenue, NYC. Digital copy available from interviewer on request via arnold.vandervalk@wur.nl.

Interview with Judith LaBelle of Glynwood on 23 May 2011, at her Cold Spring office. Digital copy available from interviewer on request via arnold.vandervalk@wur.nl.

Interview with Edie Stone and Carolin Mees of Green Thumb on 11 April 2011 at their offices. Digital copy available from interviewer on request via arnold.vandervalk@wur.nl.

Interview with Frank Vigilante, president of the community garden El Sol Brillante, located on East 12th St. in Manhattan, on 21 June 2011. Sound track available from interviewer on request via arnold.vandervalk@wur.nl.

5 Peri-urban farmland characterisation

A methodological proposal for urban planning

Esther Sanz Sanz, Claude Napoléone and Bernard Hubert

Introduction

Agriculture is currently playing a role in Western European public policy and planning by means of land-use and land-cover local regulations (Galli *et al.*, 2010; Groot *et al.*, 2009). City sprawl is encroaching onto farmland and the urban fringe of intensive agriculture that used to be an important land use is diminishing fast (EEA, 2006). Other related functions are also taking over (Delattre and Napoléone, 2011). A new set of concerns is arising, such as landscape preservation, ground-water quality, health and food security or sovereignty (Griffon, 2006). These are some of the reasons that legitimate the protection of agricultural zones in urban planning (Hervieu, 2002; Vidal and Fleury, 2009; Waldhardt *et al.*, 2010). Nowadays, agriculture's multi-functionality is credited with providing tangible benefits on ecological and economical territorial dynamics, at both local and regional levels (Groot *et al.*, 2007; Guillaumin *et al.*, 2008).

This chapter seeks to examine how local policy can include agricultural and urban planning into a singular territorial project that would be no longer urban or rural but a resilient model integrating both realms. We seek a framework to define changing agriculture practices and landscapes on urban fringes, especially those that have been gobbled up by city sprawl. By extension, we question those tools and instruments that could be developed to take into account agricultural issues in planning. Our aim is to draw up an empirical and theoretical framework, looking for a methodology to integrate peri-urban agriculture management in urban planning as a tool for policy makers and stakeholders.

Our thesis is that farming under urban influence is responding to the pressures and opportunities that arise from its geographical adjacency to cities. Five trends can be identified, based on a literature review and fieldwork:

1   An intensive and specialised high-value production farming selling both in short or long supply chains (Aubry and Kebir, 2013; Nahmias and Le Caro, 2012). These are stable farms with a long-term oriented strategy, and aware of marketing techniques.

2   Pluri-activity, low-intensive lifestyle and leisure-oriented farms, such as equine services (Busck *et al.*, 2008). Their weak cost-effectiveness is counterbalanced by an inter-sectorial strategy and by the employment of family labour force.

3   Opportunist and 'nomad' structures growing on formerly abandoned farmland close to or even encroached on by urban zones (Soulard, 2014). These are extensive production farms usually oriented by Common Agricultural Policy (CAP) primes and subventions. Their strategy is based on farmland surface increase, easily available on the urban fringe. They are frequently associated with market anticipation phenomena linked to land-use classification policy changes (Jouvé and Napoléone, 2003).

4   Independent historic farms, out of urban influence as they are protected by policy settings, such as quality growing zoning, for instance, the French *appellation d'origine contrôlée* (AOC) (Lees and Dérioz, 1994; Pérès, 2007).

5   Small farms that are non-professional and non-oriented to market (Buixade, 2009). Hence, rural farming, even under urban influence, is different from peri-urban farming that can be either focused on urban demands (Minvielle *et al.*, 2011) or take advantage of urban facilities (Hochedez and Le Gall, 2011), and it is in spatial and/or in functional interaction with urbanised zones.

A review of the abundant literature on peri-urban farming reveals that most of the monographic studies are either focused on regional cases (see for example, the French journal *Cahiers Agricultures* vol. 22(6) of 2013) or on practice-oriented studies (see the 2014 edition of the journal *Espaces et Sociétés* no. 158 dealing with 'Agriculture and City'). Furthermore, scientific knowledge provides manifold frameworks for agriculture modelling and assessment oriented to environmental evaluation (Burel and Baudry, 2004; Helming and Pérez-Soba, 2011; Schaldach and Priess, 2008; Termorshuizen and Opdam, 2009) and rural development (Groot *et al.*, 2009; Lardon, 2012; Véron, 2003; Waldhardt *et al.*, 2010; Wiggering *et al.*, 2006). Nevertheless, agriculture has been rarely considered in interaction with urbanised zones, and attempts of building a peri-urban agriculture typology are scarce. Some studies have assumed the Von Thünen model of agricultural land use that classifies farming in four rings depending on the distance to the city centre. According to the Thünensian logic of land-use distribution, adjacency to the urban market determines the profitability of agriculture, measured in terms of transport cost and locational rent, and then the location of each farming type. Nahmias and Le Caro (2012) further develop this model to define a gradient of urban agriculture based on the distance to the town but also on its functions and integration at metropolitan level. Zasada *et al.* (2013) analyse the variables characterising agriculture under urban influence to define the spatial distribution of agricultural land as well as farm structure characteristics and performance indicators in peri-urban areas. Nevertheless operational definitions considering all the issues specific to peri-urban agriculture and practices for supporting spatial management decisions and planning are lacking.

For these reasons, we seek to draw up a methodology to characterise and locate peri-urban productive agriculture, in order to construct agri-urban landscape units involving cropping systems and structures as well as urban functions and morphology, in order to integrate agriculture issues in urban planning. Our research is founded on 'resilience thinking': we first explain how resilience thinking can offer a useful framework that allows us to consider new ways of urban planning. Furthermore, the application of this concept to urban systems has been little developed in the literature. In this view, we consider that agriculture forms as part of urban systems, since cities need ecosystem services and goods provided by agricultural land. We need further to construct a spatial definition of existing peri-urban agriculture forms by means of landscape indicators. We then present the approach and methods employed for developing our methodology. We also introduce two case studies and our first results. Finally we examine how a methodology based on landscape indicators could be constructed in order to make the resilience concept operative for agri-urban planning and we introduce topics for further research.

The conceptual framework: resilience thinking and urban planning

Resilience thinking can offer a pertinent conceptual framework to understand how the city system works since it allows ecosystem functions to be integrated with socio-economic dynamics (Pickett *et al.*, 2004; Resilience Alliance, 2007). We understand resilience as the ability of a system to undergo significant fluctuations but still return to either the old or a new stable state. Multiple equilibrium states are possible. This supposes that systems can change, adapt and transform in response to stresses and strains. So then, we reject the existence of a 'normal' equilibrium state where the system should bounce back after disturbance. Contrary to planning's usual conviction that a 'good city' is the one that is in a state of equilibrium achieved by the power of the plan, we do not seek to draw a static 'ideal resilient plan'. Planning has always looked for an ideal perfect plan to implement to solve every conflict and drive society toward happiness. But we believe this is wrong. In fact, every plan and concept of ideal cities throughout urbanism's history has been unsuccessful (Choay, 1965). What is more, they have been applied in social context that were not the same as those in which they were designed, condemning them to 'solve the problems of the past' (Davoudi *et al.*, 2012).

Hence, the 'perfect equilibrium plan' not longer exists. Furthermore, a set of plans is not enough as a synthetic tool for planning. A plan fixes a static ideal image to be achieved: the equilibrium image. But as far as there exist multiple possible equilibriums, planning should be expressed by a changing, adaptable and adaptive tool. We need to go beyond planning determinism inspired by positivism and its faith in an ideal 'natural' plan (Berque, 2010). Resilience offers planning theory and practice a paradigm shift. It also offers a useful framework that allows thinking in new ways about planning since it abandons the assumptions of certainty, blueprints, forecasting and equilibrium that still persist today (Davoudi *et al.*, 2012).

We are aware that resilience thinking is not the only approach revisiting urban planning: other academic approaches seek to characterise forces acting in urban systems in order to propose new management tools. In this way, urban metabolism

(Kim and Barles, 2012) proposes a metaphorical framework to study the interactions of natural and human systems in specific regions. Their model facilitates the description and analysis of the flows of the materials and energy within cities, but does not provide clear prospective elements. The concept of the 'just city' (Fainstein, 2010) presents a model of spatial relations based on equity within the context of a global capitalist political economy. This work provides a challenging approach for evaluating urban policy and development projects from this perspective but does not consider ecological systems. Integrated planning takes a holistic and sustainable perspective mainly oriented to assessment and modelling ecological problems in urban areas (Alberti and Waddell, 2000). We base our research on resilience thinking since it offers a framework integrating both ecological and social dimensions of urban systems. Furthermore, the resilience method enables interdisciplinary work and transdisciplinary cooperation with local stakeholders with a common managed terminology.

Approach

We aim to bring scientific knowledge into decision-making about peri-urban agriculture's integration into urban planning. By extension, we question the tools and instruments that could be developed to take into account agricultural issues. But interactions between urban and agricultural systems are so strong that we cannot any longer consider them as individual systems, but as one complex one (Fleury *et al.*, 2003), in other words, the '*campagnes urbaines*' (Donadieu, 1998):

> Agricultural landscapes, which occupy more than 40% of the European Union and more than 30% of the contiguous American states, are arguably the largest urban land use, since the functional ecosystems of cities extend to agricultural watersheds that provide potable water and other ecosystem services, and the supply chains of urban food processing and consumption begin in agricultural landscapes.
>
> (Nassauer, 2012: 223)

We need to build a systemic, generic and operational spatial definition of existing forms of agriculture on the urban fringe by the means of measurable and qualitative landscape indicators in order to support local planning. How can peri-urban agriculture be characterised without relying on definitions based on distance to city centres? We propose a methodology founded on a triple approach to agricultural issues: 1) farming area characterisation (landscape structures); 2) socio-economic analysis of farming activities (landscape functions); and 3) land-use and land-cover policy settings (landscape policies).

For this purpose, we adopt an agronomic and geographical approach (Benoît *et al.*, 2012; Deffontaines *et al.*, 1995; Deffontaines and Thinon, 2006, 2008; Hubert *et al.*, 2004; Rizzo *et al.*, 2013; Thinon *et al.*, 2007) combining a morphological analysis (landscape spatial patterns) with an analysis of *geographical fields* (the spatial attributes affecting land-use localisation on peri-urban areas) by the means of remote sensing interpretation, field surveys, onsite landscape reading and statistical analysis.

Case studies

We test our methodological proposal in two case studies: Madrid metropolitan area (Spain) and Avignon urban area (France). Both regions have different urban sprawl models (compact versus scattered) as well as different policy settings. On the one hand, Madrid is a metropolis with compact and dense development. The study area has a population of more than 4 million people over 842 km² with an average density of 4,930 inhabitants/km². On the other hand, Avignon is a low-density dispersed urban area. The urban area of Avignon, as defined by the Institut National de la Statistique et des Etudes Economiques (National Institute of Statistics and Economic Studies) (INSEE) in 2010, is the least dense urban area in France (323 inhabitants/km² in Avignon but an average of 820/km² for French urban centres), with little difference between the urban centre and its surroundings (1 to 3.5, according to AURAV, 2012). For our purpose, we have delimited a sample of 385km² on the Vauclusien side of Avignon urban area, with a density of 518 inhabitants/km². We decided to test our methodology in such different cases in order to verify its capacity to be systemic and generic, testing that it could be transposed to any other city and study case.

We decided to apply the resilience framework in a territory defined by operational administrative boundaries because our research is oriented to policy makers and planners. in France and in Spain, the inter-municipality and municipality scales are the levels where decisions concerning peri-urban land use are taken. They correspond to the scale where agri-urban projects could be implemented. Delimitation of the study areas was based on two criteria.

First, administrative boundaries were used. Each study area must overlap the administrative boundaries of a municipality or a group of associated municipalities,[1] preferably possessing a (peri-urban) agriculture policy. In the Avignon case (see Figure 5.1), we took as a starting area the demarcation of the urban area of Avignon defined by INSEE, to focus then on the perimeter of the Communauté de Communes du Pays de Sorgues-Monts de Vaucluse and on the Vauclusien side of the Communauté de Communes du Grand Avignon.

In the Madrid case (see Figure 5.2), we started wth the demarcation (non statuary) of the metropolitan area of Madrid. We chose study one municipality of the urban or inner ring (Fuenlabrada), one municipality of the metropolitan area or outer ring (Rivas Vaciamadrid) and, finally, one municipality of the metropolitan region or Gran Madrid (Morata de Tajuña).

Second, natural or built geographical boundaries were used. Each study area is considered as a geographical unit with a specific and singular landscape. In both situations, the study areas are representative of the historical agricultural landscape in the urban or metropolitan context. Rivers, mountains or motorways limit them. In the Avignon case, it is part of the Sorgues watershed, defined by the Rhône river at West, the Durance river and Avignon road (D900) to the south, the Vaucluse

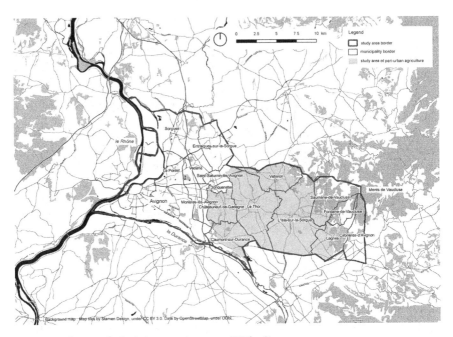

Figure 5.1 Case study in Avignon urban area (385km²)

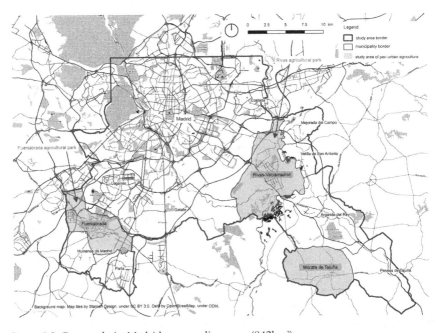

Figure 5.2 Case study in Madrid metropolitan area (842km²)

plateau mountains in the east and the river Sorgue d'Entreaigues and the roads D28and D6 to Sorgues to the north. It is a sample of the regional landscape type *Le pays des Sorgues*,[2] a sub-unit of the *La plaine Comtadine* landscape type i.e. agriculture irrigated by the means of a canal network designed on former drained swamps, farming plots and separated by hedges. The study area also includes Avignon city and its urban cluster development, representatives of the *Le couloir Rhodanien* landscape type. In the Madrid case, there is a sample of the regional landscape unit *Vegas del Jarama y Tajuña*[3] on the southeastern side of the metropolitan area of Madrid and of Gran Madrid. That is, irrigated agriculture by the means of a canal and dam network, along the A3 highway in the Jarama and Tajuña watersheds. A sample of the regional landscape unit *Campiñas metropolitanas* lies to the south-west and is irrigated agriculture by the means of old wells (the traditional canal network does not exist here) in the metropolitan area of Madrid along road A42.

Methods

Remote sensing interpretation

We carried out a diachronic analysis of satellite images from both study areas for the the 1970s, the late 1980s and today. This time sequence covers the relevant periods of change in agri-urban regimes (Berque, 2010; Grosso *et al.*, 1993; Naredo, 2010; Naredo and García Zaldívar, 2008; Wiel, 1999):

- 1970s: rapid urban expansion begins, facilitated by the automobile boom. Historical vegetable cropping is pushed back by the first ring of urbanisation around city.
- Late 1980s: consolidation of the commercial areas and residential urban expansion around the city i.e. second ring urbanisation.
- Today: unmanaged, widespread urban growth.

The principal events affecting the transformations of agri-urban systems are summarised in Figure 5.3.

We analysed land-use changes of five land classes: urban (buildings and infrastructures), agricultural, natural (woods and unconstructed slopes), bare soils and water bodies, aiming to examine the spatial evolution of both urban and agricultural land to determine how the urban growth model influences agriculture typologies in peri-urban contexts.

Field surveys

We carried out two field surveys in Madrid in June 2014 (two weeks) and June 2015 (three weeks). Interviews were carried out with 19 farmers, 4 experts and 5 local managers. The field survey in Avignon was carried during autumn 2014 and is still in progress. To date, 13 interviews have been carried out with farmers and

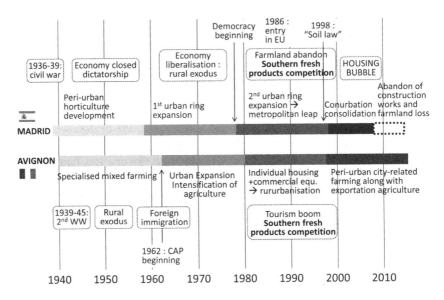

Figure 5.3 Timeline of changes in agri-urban regimes

two with experts. The Avignon fieldwork is longer than in Madrid because the agricultural issue is more complex. We aim to understand farms transformations as well as to understand farmers' strategies and their determinants.

Onsite landscape reading

We are also conducting onsite readings to recognise and trace existing crops at plot scale. Our aim is to analyse crop changes to understand peri-urban agriculture adaptation to each singular context and external drivers, for example in response to changes in local land-use plans.

Analysis

First, we need to delimit the spaces of peri-urban agriculture in order to charac-terise them. According to our fieldwork and literature review, we can identify a preliminary brief agriculture typology on the urban fringe: a) urban agriculture (UA): farming close to or into the city that endures because it is situated on a protected (frequently environmental) delimitated area and takes advantage of the proximity to the city centre's high concentration of consumers; b) rural agriculture (RA): farming far from the city influence that bases its strategy on agronomic and commercialisation criteria, including urban market-oriented farms benefiting from urban rents; c) peri-urban agriculture (PUA): uncertain farming with varied and often contrasting strategies, going from the opportunist big farms with annual crops benefiting from CAP payments, to equine services.

For our purpose, we focusing on the last type: peri-urban agriculture. We locate it, according to our fieldwork and literature review concerning urban systems, considering first the distance-time between the inner urban area and the peri-urban ring (see Figure 5.4a) and then the distance-time between the area of work/business concentration to the workers' residential area, that is the outer peri-urban villages (see Figure 5.4b). The values of these distance-times are specific to each urban area. For Avignon, according to INSEE, these values are 30 and 45 minutes respectively. The spaces of peri-urban agriculture are so located in between the peri-urban commuting areas delimitated by these distances-times (see Figure 5.4c).

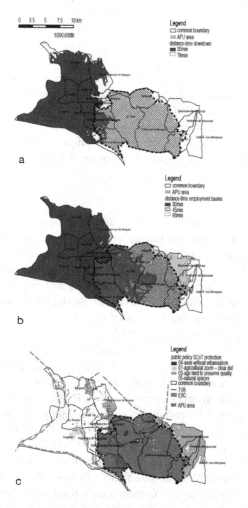

Figure 5.4 Delimitation of the spaces of peri-urban agriculture, showing (a) distance-time to the inner urban area, (b) distance-time to work concentration area, and (c) public regulations

Second, we characterise peri-urban agriculture by combining quantitative and qualitative variables (see Figure 5.5). We define peri-urban agriculture landscape units (APUs), which are intermediary tools to help planners to read agriculture. These variables come from different disciplines: geography (accessibility, altitude, etc.), agronomy (irrigation, farm size, fields arrangement, culture system, etc.), economy (land tenure, CAP subventions, etc.) and political organisation (environmental protection zones, land-use plans, etc.). We followed two steps: first, we 'manually' defined and located peri-urban agriculture landscape units in the study areas by spatialising the variables by means of Geographic Information System (GIS) software; then, we systemised the methodology by a statistical approach to be further applied.

Our analysis is founded on a triple approach to agricultural issues.

Landscape structures

We seek to characterise peri-urban farming areas. Despite the small size of the case study areas, we identified a wide range of farms, concerning their size, distribution of farming plots and cropping systems, from little family farms of barely more than 1ha of *huerta* or poultry farms, to great farms of 5,000ha with plots dispersed on different municipalities. The proportion of professional farmers varies equally depending on the APU and generally decreases in the APUs imbricated with the urban fabric and in those areas that were already quite built up the 1980s.

Landscape functions

According to the analysis of farming practices, crops are diverse depending on biophysical conditions, market opportunities and farmer strategy. We found that vegetable crops, orchard and feed grain are located on irrigated land. Cereals, olive groves, vineyards and livestock are located instead on land that is not irrigated. But we also found cereals on irrigable land that were formerly cultivated with vegetables or fruit arboriculture, and that are no longer being irrigated. Product diversity is also a function of the determinants above. Small farms (<4ha) are usually high diversified (*huerta* and poultry farms), while big farms (>100ha) are mono-oriented (cereal, vineyard, olive grove) or working with two non-irrigated crops (cereal-olives; vineyard-olives). In this way, medium-sized farms (4–30ha) can be either mono-oriented (garlic; chard) or diversified (wide range of vegetables or fruits, usually both; sometimes combined with livestock), or be oriented to three or four irrigated crops (vegetables both in open fields and in green houses; fruit arboriculture). In the French case study, larger medium-size farms (15–25ha) are sometimes cultivating some hectares of *Appellation d'Origine Contrôlée* (AOC)[4] certified vineyard in plots relatively far from the farmstead in order to boost their income, as long as AOC wine is sold easily and well. Medium-size farms and especially small farms are more likely to be organic, either due to farmers' convictions or market opportunities. The motivations to change to organic farming and the strategies of organic farms have been further studied elsewhere (see for example Lamine and Bellon, 2009).

Landscape policies

Considering land-use and land-cover policy settings, the most relevant emerging determinant is urban pressure and the probability of land becoming developable. Currently, in Spain and France, a land plot can be developed and built upon only if the municipal land use plan allows it and if it is not included in a natural or farm-land preservation area. We noticed that farms located in protected delimitated areas close to the city persist as they take advantage of the urban concentration of consumers and customers while resisting urban pressure. However, those farms that are waiting for a change in land-use planning to sell their land for building are developing opportunistic and short-term strategies not related to agronomic or market criteria. When zoning is the primary tool of land-use regulation, land-use conversion anticipation influences agricultural land prices. Potentially developable land prices are far higher than the price of land where agriculture is the only use permitted (Géniaux *et al.*, 2011). Farmland consumption for developable areas seems to be less likely when farms are more profitable than the median profitabil-ity for the sector (Chanel *et al.*, 2014).

Therefore, the APUs define spatially homogenous areas with similar character-istics in terms of landscape patterns and farming uses (see Figure 5.5). They are bigger than farms; their areas are between 20 and 40km^2. Each APU also has the same level of 'incertitude', which is the risk of farming land being abandoned due to urban pressure and perceived development possibilities. In this way, APU cartography (see Figure 5.6) can guide public action to focus on the areas with higher urban pressure. The characterisation of peri-urban agriculture allows public action to be adapted depending on the incertitude. Each APU conveys specific planning concerns and urgency, and it should be related to precise public policies. Landscape planning tools should be proposed to integrate agriculture management in urban planning and enable adaptive territorial management. Finally, since the APUs are statistically defined objects, the methodology proposed is systemic and generic and can be easily used and transposed to any city.

Conclusion and discussion

The city seems to influence most peri-urban farms, or at least to have an impact on farmers' lifestyle. Urban concentration of food consumers and customers and other activities generates new and promising opportunities of specialisation, rather than anticipation and short-term positions. Farm strategy is in most cases the result of spontaneous decisions depending on the farmer's own situation, European agri-culture policy, commercialisation options, labour availability and farmer expertise.

Scientific knowledge provides different frameworks for agriculture assessment oriented to environmental evaluation and rural development. But rural agriculture is different from urban agriculture that is focused on urban needs (Fleury *et al.*, 2003). Empirical knowledge has been reported concerning peri-urban agriculture integration within the urban planning project focus on agriculture as a green facil-ity (Galli *et al.*, 2010; Poulot, 2011). Nevertheless, integrated frameworks are lacking

landscape structures

urban spatial configuration

relation with urban zones	accessibility (road network)	nearby urban fabric morphology	land-use evolution (in 2011, 1987, 1973, 1945)	link to farmstead
(i) imbrication	_checked /radial pattern	(1) continuous urban fabric	(0) non built	(M) around farmstead
(j) juxtaposition	_dense / non-dense	(2) discontinuous urban fabric	(1) slightly built	(S) non nearby farmstead
(c) little contact	_regular / non-regular	(3) non-dense scattered settlements	(2) a bit built	
(p) no contact		(4) isolated settlements	(3) quite built	
			(4) quite built imbricated clusters	

farming spatial configuration

natural resources
_altitude
_steep slope & exposure

irrigation
(i) irrigated
(a) irrigable but non irrigated
(s) dryland

farm/field size	fields shape	fields arrangement	crops	crop transitions
_no. fields/farm	(C) rectangular/regular	(1) grouped	_% arable crops surface	% perennial --> vegetables
_farm size median [ha]	(M) crooked	(2) nearby	_% vegetables crops surf.	% perennial --> gardening
_field size median[ha]	(I) irregular	(3) scattered aggregates	_% perennial crops surface	% vegetables --> arable crops
_farm density/local average		(4) fragmented farm	_% viticulture surface	% garden c. --> arable crops
_field density/local average			_% livestock & feed crop surf.	% perennial --> arable crops
			_% NA surface	% perennial --> livestock
				% perennial --> equine

landscape fonctions

farming practices

professional farming	farming dynamics	farm economic orientation		farmland tenure system
_% no. professional farming	_% no. recently created farms	_% no. mono-cropping farm		(f) tenant farming
_% no. farms >SMI	_% no. recently shut up farms	_% no. crop-diversity farms		(d) owwner-farmed holding
	_% no. recently extended farms	**specialized farm economic orient.**		
	_% no. recently reduced farms	% market gardening		
	_% surface cultivated formerly abandoned	% garden centre		
	_% surface recently abandoned	% greenhouse		
		% quality wine (AOC)		
		% equine		

landscape policies

preservation policies & regulations

preservation regulations	local land use classification
_% PNR (environmental, parks)	_% "agricultural land"
_% PPRI (flood risk)	_% "natural land"
_% AOC (quality growing zoning)	_% "developable land"
_% PLU (local urban plan)	
_% ZAP, PAEN... (farmland protec.)	

Figure 5.5 Peri-urban agriculture characterisation: variables

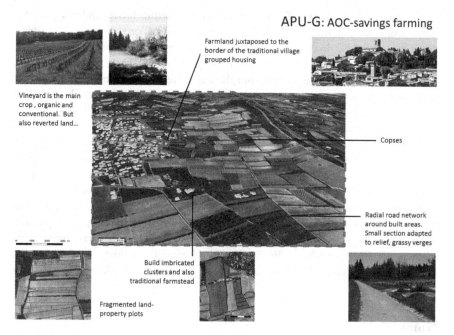

APU-G: AOC-savings farming

Farmland juxtaposed to the border of the traditional village grouped housing

Vineyard is the main crop, organic and conventional. But also reverted land...

Copses

Radial road network around built areas. Small section adapted to relief, grassy verges

Build imbricated clusters and also traditional farmstead

Fragmented land-property plots

Figure 5.6 Example of a graphic summary sheet representing the main features of an APU oriented to patrimonial farming linked to quality wine production

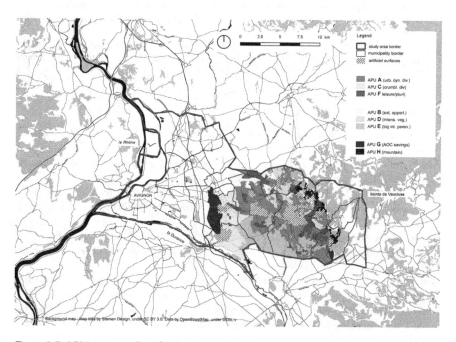

Figure 5.7 APU cartography of Avignon case study

that consider all issues specific to peri-urban agriculture forms and practices for supporting spatial management decisions and planning. For this purpose, we have drawn up a methodology to characterise and assess peri-urban productive agriculture in order to construct peri-urban landscape units involving cropping systems and structures as well as urban morphology. We are currently applying this methodology to two study cases in France and Spain. This practical work will allow the model to be further improved so it can be easily applied elsewhere as a tool for policy makers and stakeholders.

Moreover, it would be interesting to undertake further research to relate the peri-urban agriculture typologies to urban growth patterns in order to identify possible correlations. Finally, landscape planning tools should be proposed to integrate agriculture management in urban planning (Vidal and Fleury, 2009) and hence enable hence adaptive territorial management. We can imagine proactive public policies better involving farming activities in urban development projects going beyond demarcation protection, to integrate farming as an activity complementary to urban activities. We need to go beyond the reduction of the city as urban and agriculture as rural to think about urban and agricultural territory as a complex system.

Notes

1   In France this administrative level is called a *Communauté de Communes.*
2   Landscape type defined thus in Agence Paysages *et al.*, 2013.
3   Landscape types defined thus in several publications: Martinez Garrido and Mata Olmo, 1987; Mata Olmo and Mato Miguel, 2010; Mata Olmo and Rodriguez Chumillas, 1987.
4   The *Appellation d'Origine Contrôlée* (AOC) translates as 'controlled designation of origin', and is the French certification granted to certain wines and other agricultural products.

References

Agence Paysages, DIREN, DDE, & Conséil Général Vaucluse. (2013). *Atlas des paysages de Vaucluse.* Conseil Général de Vaucluse – DREAL PACA.
Alberti, M., & Waddell, P. (2000). An integrated urban development and ecological simulation model. *Integrated Assessment, 1*(3), 215–227.
Aubry, C., & Kebir, L. (2013). Shortening food supply chains: A means for maintaining agriculture close to urban areas? The case of the French metropolitan area of Paris. *Food Policy, 41*, 85–93.
Benoît, M., Rizzo, D., Marraccini, E., Moonen, A. C., Galli, M., Lardon, S., … Bonari, E. (2012). Landscape agronomy: a new field for addressing agricultural landscape dynamics. *Landscape Ecology, 27*(10), 1385–1394.
Berque, A. (2010). *Milieu et identité humaine. Notes pour un dépassement de la modernité.* Paris: Donner lieu.
Buixade, I. A. (2009). Les dificultats de manteniment de l'agricultura periurbana. L'exemple de l'horta de Lleida. *Scripta Nova – Revista Electrónica de Geografía Y Ciencias Sociales, 13*(284), 1–31.

Burel, F., & Baudry, J. (2004). *Landscape ecology: concepts, methods, and applications.* Enfield, NH [u.a.]: Science Publ.

Busck, A. G., Kristensen, S. P., Præstholm, S., & Primdahl, J. (2008). Porous landscapes – The case of Greater Copenhagen. *Urban Forestry & Urban Greening,* 7(3), 145–156.

Chanel, O., Delattre, L., & Napoléone, C. (2014). Determinants of Local Public Policies for Farmland Preservation and Urban Expansion: a French Illustration. *Land Economics, 90*(3), 411–433.

Choay, F. (1965). *L'Urbanisme, utopies et réalités: une anthologie.* Paris: Seuil.

Davoudi, S., Shaw, K., Haider, L. J., Quinlan, A. E., Peterson, G. D., Wilkinson, C., ... Davoudi, S. (2012). Resilience: A bridging concept or a dead end? 'Reframing' resilience: Challenges for planning theory and practice interacting traps: Resilience assessment of a pasture management system in northern Afghanistan urban resilience: What does it mean in planning practice? Resilience as a useful concept for climate change Adaptation? The politics of resilience for planning: A cautionary note. *Planning Theory & Practice, 13*(2), 299–333.

Deffontaines, J. P., & Thinon, P. (2006). Emergence d'un concept: Un itinéraire entre agronomie et géographie. In J.-M. Legay (ed.), *L'interdisciplinarité dans les sciences de la vie,* 45–50. Éditions Quae.

Deffontaines, J.-P., & Thinon, P. (2008). La cartographie d'unités agro-physionomiques. Analyser la répartition et la dynamique des usages agricoles dans le territoire. *Résultats Des Recherches Du Département INRA – SAD,* (27), 1–4.

Deffontaines, J. P., Thenail, C., & Baudry, J. (1995). Agricultural systems and landscape patterns: how can we build a relationship? *Landscape and Urban Planning, 31*(1–3), 3–10.

Delattre, L., & Napoléone, C. (2011). Écologiser les documents d'urbanisme pour protéger les terres agricoles et les espaces naturels. *Courrier de L'environnement de l'INRA,* (60), 63–75.

Donadieu, P. (1998). *Campagnes urbaines.* Arles/Rennes: Actes Sud /Ecole nationale supérieure du paysage.

EEA (European Environment Agency). (2006). *Urban sprawl in Europe: the ignored challenge.* Copenhagen, Denmark; Luxembourg: European Environment Agency.

Fainstein, S. S. (2010). *The Just City.* New York: Cornell University Press.

Fleury, A., Moustier, P., & Tolron, J.-J. (2003). Multifonctionnalité de l'agriculture dans les territoires périurbains : diversité de formes d'exercice du métier d'agriculteur, insertions de l'agriculture dans l'aménagement des territoires. *Les Cahiers de La Multifonctionnalité,* (2), 83–91.

Galli, M., Lardon, S., Marraccini, E., & Bonari, E. (2010). Agricultural management in peri-urban areas. In *The experience of an international workshop. Land Lab–Scuola Superiore Sant'Anna (Italy), INRA et AgroParisTech-ENGREF, UMR Métafort Clermont Ferrand (France).* Ghezzano, Italy: Felici Editore.

Géniaux, G., Ay, J.-S., & Napoléone, C. (2011). A spatial hedonic approach on land use change anticipations. *Journal of Regional Science, 51*(5), 967–986.

Griffon, M. (2006). *Nourrir la planète: pour une révolution doublement verte.* Paris: Odile Jacob.

Groot, J. C. J., Rossing, W. A. H., Jellema, A., Stobbelaar, D. J., Renting, H., & Van Ittersum, M. K. (2007). Exploring multi-scale trade-offs between nature conservation, agricultural profits and landscape quality—A methodology to support discussions on land-use perspectives. *Agriculture, Ecosystems & Environment, 120*(1), 58–69.

Groot, J. C. J., Rossing, W. A. H., Tichit, M., Turpin, N., Jellema, A., Baudry, J., ... van de Ven, G. W. J. (2009). On the contribution of modelling to multifunctional agriculture: Learning from comparisons. *Journal of Environmental Management, 90,* 147–160.

Grosso, R., Galas, J., Locci, J.-P., & Clap, S. (1993). *Histoire de Vaucluse. Volume 2: Les Vauclusiens, des campagnes à la ville.* Avignon (France): Éditions A. Barthélemy.

Guillaumin, A., Dockès, A.-C., Tchakérian, E., Daridan, D., Gallot, S., Hennion, B., ... Perrot, C. (2008). Demandes de la société et multifonctionnalité de l'agriculture: attitudes et pratiques des agriculteurs. *Courrier de L'environnement de l'INRA,* (56), 45–66.

Helming, K., & Pérez-Soba, M. (2011). Landscape scenarios and multifunctionality: making land use impact assessment operational. *Ecology and Society, 16*(1), 50.

Hervieu, B. (2002). La multifonctionnalité de l'agriculture : genèse et fondements d'une nouvelle approche conceptuelle de l'activité agricole. *Agricultures, 11*(6), 415–419.

Hochedez C., & Le Gall, J. (2011). Nord et Sud face aux crises. De nouveaux réseaux maraîchers métropolitains au service d'une agriculture de proximité: les cas de Buenos Aires et Stockholm. *Norois, 221*(4): 25–38.

Hubert, B., Moulin, C.-H., Roche, B., Pluvinage, J., & Deffontaines, J.-P. (2004). Quels dispositifs pour conduire des recherches en partenariat ? L'intervention d'une équipe de recherche au Pays basque intérieur. *Économie Rurale, 279*(1), 33–52.

Kim, E., & Barles, S. (2012). The energy consumption of Paris and its supply areas from the eighteenth century to the present. *Regional Environmental Change, 12*(2), 295–310.

Jouvé, A. M., & Napoléone, C. (2003). Stratégies des agriculteurs et réorganisations spatiales sous contrainte de la périurbanité: étude du pays d'Aix-en-Provence, In *Bouleversements fonciers en Méditerranée. Des agricultures sous le choc de l'urbanisation et des privatisations,* 143–172. Paris: Karthala/CIHEAM.

Lamine, C., & Bellon, S. (2009). Conversion to organic farming: a multidimensional research object at the crossroads of agricultural and social sciences. A review. *Agronomy for Sustainable Development, 29*(1), 97–112.

Lardon, S. (ed.). (2012). *Géoagronomie, paysage et projets de territoire. Sur les traces de Jean-Pierre Deffontaines.* Editions Quae – NSS Dialogues.

Lees, C., & Dérioz, P. (1994). Le jardin de la France au péril de la ville: place et évolution de l'activité agricole dans le Grand Avignon (The garden of France at the city's peril: place and evolution of agricultural activity in Greater Avignon). *Bulletin de l'Association de géographes français, 71*(2), 170–180.

Martinez Garrido, E., & Mata Olmo, R. (1987). Estructuras y estrategias productivas del regadío metropolitano de Madrid. *Agricultura y Sociedad,* (42), 149–180.

Mata Olmo, R., & Mato Miguel, J. F. (2010). Los regadíos históricos del Tajuña. Río Tajo. In *Los regadíos históricos españoles. Paisajes culturales, paisajes sostenibles.* Madrid: Ministerio de Medio Ambiente y Medio Rural y Marino.

Mata Olmo, R., & Rodriguez Chumillas, I. (1987). Propiedad y explotación agrarias en el regadío de las 'vegas' de Madrid. *Agricultura y Sociedad,* (42), 149–180.

Minvielle, P., Consales, J.-N., & Daligaux, J. (2011). Région PACA: le système AMAP, l'émergence d'un SYAL métropolitain. *Économie Rurale, 2*(322), 50–63.

Nahmias, P., & Le Caro, Y. (2012). Pour une définition de l'agriculture urbaine: réciprocité fonctionnelle et diversité des formes spatiales. *Environnement Urbain/Urban Environment,* (6), 1–16.

Naredo, J. M. (2010). *Presión inmobiliaria y destrucción de sistemas agrarios y suelos de calidad. El ejemplo de la Comunidad de Madrid.* Madrid: Sociedad Española de Historia Agraria – Documentos de Trabajo, 10(04).

Naredo, J. M. & García Zaldívar, R. (coords) (2008). *Estudio sobre la ocupación de suelo por usos urbanos-industriales, aplicado a la Comunidad de Madrid.* Madrid: Universidad Politécnica de Madrid.

Nassauer, J. I. (2012). Landscape as medium and method for synthesis in urban ecological design. *Landscape and Urban Planning*, *106*(3), 221–229.

Pérès, S. (2007). *La vigne et la ville: forme urbaine et usage des sols*. Université Montesquieu-Bordeaux IV.

Pickett, S. T. A., Cadenasso, M. L., & Grove, J. M. (2004). Resilient cities: meaning, models, and metaphor for integrating the ecological, socio-economic, and planning realms. *Landscape and Urban Planning*, *69*(4), 369–384.

Poulot, M. (2011). Des arrangements autour de l'agriculture en périurbain: du lotissement agricole au projet de territoire: Exemples franciliens. *VertigO*, *11*(2). Online: http://vertigo.revues.org/11188 [accessed 3 July 2012].

Resilience Alliance. (2007). *Research Prospectus: A Resilience Alliance Initiative for Transitioning Urban Systems towards Sustainable Futures [on line]*. CSIRO (Australia), Arizona State University (USA), Stockholm University (Sweden).

Rizzo, D., Marraccini, E., Lardon, S., Rapey, H., Debolini, M., Benoît, M., & Thenail, C. (2013). Farming systems designing landscapes: land management units at the interface between agronomy and geography. *Geografisk Tidsskrift-Danish Journal of Geography*, *113*(2), 71–86.

Schaldach, R., & Priess, J. A. (2008). Integrated models of the land system: a review of modelling approaches on the regional to global scale. *Living Reviews in Landscape Research*, *2*(1). Online: www.livingreviews.org/lrlr-2008-1 [accessed 02 May 2013].

Soulard, C.-T. (2014). Les agricultures nomades, une caractéristique du périurbain. *Pour*, *224*(4), 151.

Termorshuizen, J. W., & Opdam, P. (2009). Landscape services as a bridge between landscape ecology and sustainable development. *Landscape Ecology*, *24*(8), 1037–1052.

Thinon, P., Martignac, C., Metzger, P., & Cheylan, J.-P. (2007). Analyse géographique et modélisation des dynamiques d'urbanisation à La Réunion. *Cybergeo*, *16*(389). Online: http://cybergeo.revues.org/8692 [accessed 1 March 2016].

Véron, F. (ed.). (2003). *Les cahiers de la multifonctionnalité n.2. Propositions de recherche soutenues par le dispositif INRA-CEMAGREF-CIRAD sur la multifonctionnalité de l'agriculture et des espaces ruraux*. INRA Editions.

Vidal, R., & Fleury, A. (2009). La place de l'agriculture dans la métropole verte. *Projets de Paysage*. Online: www.projetsdepaysage.fr/fr/la_place_de_l_agriculture_dans_la_metropole_verte [accessed 1 March 2016].

Waldhardt, R., Bach, M., Borresch, R., Breuer, L., Diekötter, T., Frede, H.-G., & Julich, S. (2010). Evaluating today's landscape multifunctionality and providing an alternative future: a normative scenario approach. *Ecology and Society*, *15*(3), 30.

Wiel, M. (1999). *La transition urbaine, ou, Le passage de la ville-pedestre a la ville-motorisee*. Sprimont (Belgium): P. Mardaga.

Wiggering, H., Dalchow, C., Glemnitz, M., Helming, K., Müller, K., Schultz, A., … Zander, P. (2006). Indicators for multifunctional land use – Linking socio-economic requirements with landscape potentials. *Ecological Indicators*, *6*(1), 238–249.

Zasada, I., Loibl, W., Köstl, M., & Piorr, A. (2013). Agriculture under human influence: A spatial analysis of farming systems and land sue in European rural-urban-regions. *European Countryside*, *5*(1), 71–88.

6 Garden cities

A flawed model for ecological urbanism?

Greg Keeffe, Natalie A. Hall and Andy Jenkins

Introduction

In his book, *Garden Cities of To-Morrow*, Ebenezer Howard (1902 [2014]) discusses the 'devastating consequences' of urbanization, of people continuing to stream into already overcrowded cities, and thus further depleting the country districts. Today, the issue is not so much one of depleting country districts but one of depleting global resources and the Garden City model is still regarded by many as a solution. As of 2010, more than half of the global population was living in cities (World Health Organization, 2014). These growing cities concentrate large proportions of the middle- and upper-income populations, which results in the convergence of demand for goods and services at a single location. This ever-growing demand has caused cities to become increasingly reliant on drawing resources from across the globe. Urban populations account for enormous flows of energy, materials and water, while dispersing wastes with a negative impact on the environment (Ferrao and Fernandez, 2013). Over a century since *Garden Cities of To-Morrow* was first published, Ferrao and Fernandez (2103) echo Howard's concerns and argue that urban centres reflect the societal stresses associated with the rapid changes in the world economy, as illustrated by the fact that, while most of the worlds economic activities are concentrated in cities, producing wealth and improving education and access to markets, a significant proportion of the world's population with unmet needs continues to live in urban areas (Ferrao and Fernandez, 2013). This is very much an international issue as supported by figures released by Oxfam in 2013 that showed over 500,000 people in the UK did not have sufficient access to food (Cooper and Dumpleton, 2013). Newman and Jennings (2008) suggest that as cities grow in population, the tendency is to expand in land area, consuming important natural ecosystems and agricultural land. Carolyn Steel's, *Hungry City* warns the UK has long since exceeded its productive capacity, with London alone currently requiring a combined area more than a hundred times larger than itself to feed it, roughly equivalent to all of the productive farmland in the UK (Steel, 2009). The Garden City model has been widely accepted around the world, with many new towns and suburbs being established through the application of Howard's proposal (see www.garden-cities-exhibition.com). Though devised in the late 17th century, many blindly cite Howard's model as a solution for the issues facing modern UK cities.

As recently as April 2014, a press conference at Westminster saw the then-Deputy Prime Minister Nick Clegg insist 'Planned garden cities can help tackle the need for housing in southern England without inflicting the same level of damage on the countryside as urban sprawl'. Prime Minister David Cameron described his vision for future communities as 'green, planned, secured with gardens' (*The Independent*, 7 November, 2013). It could be argued that recent plans for the creation of new Garden Cities in the UK seem to be more focused on a nostalgic architectural aesthetic than the principles originally set out by Howard, often omitting local agriculture, a prominent feature of Howard's design. It could be argued that Howard's proposal of the Garden City is a more sustainable model than what is currently being suggested under the same label. This chapter aims to explore Howard's original proposal and using an analysis of the metrics proposed, establish whether the Garden City is indeed a valid model for ecological urbanism.

Introduction to the Garden City model

London-born Ebenezer Howard spent most of his early adulthood in America, having initially left the UK to pursue farming in Nebraska. Following the failure of his venture, Howard relocated to Chicago where he stayed for four years. During his time there, Chicago was experiencing rapid growth and as a consequence, terrible housing shortages. The response to the expansion crisis would see Chicago becoming known for its 'planned landscaping', a 'blend of town and country' (International Garden Cities Exhibition, 2014). Upon his return to London, Howard recognized many of the issues resulting from rapid urban expansion. By 1898 Howard had formalized his ideas on how to resolve the many problems generated by urbanization and as a result published his book, *To-Morrow: A peaceful path to reform*, later re-issued in 1902 as *Garden Cities of To-Morrow*. The mass migration of rural populations to cities was causing 'smog-riden slums' and overcrowding in cities while also leaving the countryside desperate for labour, decreasing productivity a result of the 'land left idle' (International Garden Cities Exhibition, 2014). Howard discussed the UK's industrial paradigm, which saw 'sharp lines divide agriculture from industry' and described it as 'universally considered' to be an enduring one. He argued that this was a common 'fallacy' which ignored the possibility of alternatives. Howard believed the key to stemming the negative effects of urbanization was to create new cities 'elsewhere', cities that would combine 'all the advantages of the most energetic and active town life, with all the beauty and delight of the country'. In his book, Howard outlined his proposal of the Garden City.

The Garden City concept is most often described as a 'self-sufficient entity' capable of accommodating a stable population, and 'each ringed by an agricultural belt' (Brittanica Encyclopedia, 2014). Howard himself described a 6000 acre (2428ha) area with the potential to house a population of 32,000. Howard described how the 6000 acres, exhibiting a 'possible circular form' (Howard, 1902 [2014]) would be divided into zones, with a central 'city' accounting for one sixth of the total area, approximately 1000 acres or 405ha and accommodating 30,000 residents, with the remaining 2000 residents occupying the 5000 acre or 2023ha agricultural estate. Howard's circular design would center around a 'five and a half acre (2.2ha) beautiful and well-watered garden' (Howard, 1902 [2014]). The various functions within the city such as industry, residential, recreational and retail were again divided into zones and arranged in concentric circles radiating from the central garden; each differing in size according to the function of the individual zone. Included within these radiating zones is what Howard referred to as the 'Grand Avenue', at 420ft (128m) in width, Howard described a band of green that 'divides the part of town which lies outside of central Park into two belts', though he admits 'it really constitutes an additional park of 115 acres'.

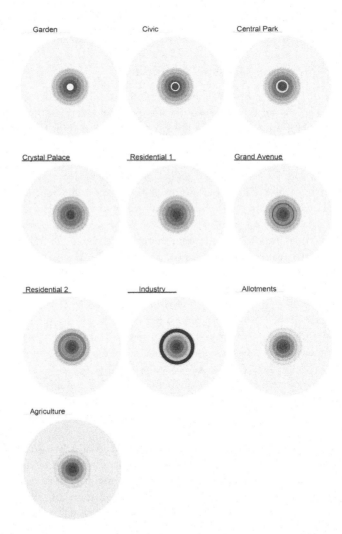

Figure 6.1 Illustration highlighting the multiple zones used within Howard's model

The entire 6000 acre city would then be dissected into six 'wards' by what Howard described as six 'magnificent boulevards'. Howard suggested these boulevards would be 120ft (36.5m) in width and span from the centre garden to the outer limits of the city. The outer most ring of the central city would be reserved for industry, and within this zone Howard describes how 'factories, warehouses, dairies, markets, coal yards and timber yards etc.' face onto the 'circle railway'. The circle railway follows the perimeter of the city with sidings that connect it to the 'main line': a larger railway connecting one Garden City to the next. Howard suggests that this arrangement would have positive impacts on the local industries as it would allow for fast and efficient distribution of goods while also presenting 'great savings in regard to packing and cartage', as well as minimizing 'loss from breakages'. Localized production and consumption were encouraged by Howards Garden City model, he emphasized: 'There are 30,000 townspeople to feed' and 'every farmer now has a market close to his doors' (Howard, 1902 [2014]), and while local farmers may not be able to supply products, such as tea, coffee or tropical fruits, there was a 'beam of hope' for local farmers in the connectivity and proximity offered by the Garden City model. Indeed, 'while the American has to pay railway charges to the seaboard, charges for Atlantic transit and railway charges to the consumer, the farmer of Garden City has a market at his very doors' (Howard, 1902 [2014]). Howard also highlighted the decline of fruits and vegetables grown in the UK, a trend that has continued with time, and he argued the cause of this decline was owing to 'the high charges for freights and commission'. Howard did however find exception to this trend, claiming 'Farmers, except near towns', reinforcing his argument on the importance of a local market.

Howard's approach to waste treatment within the Garden City could be viewed as a basic example of closed loop urbanism: 'The refuse of the town is utilized on the agricultural portions of the estate' (Howard, 1902 [2014]). He argued the effectiveness of this method of recycling by suggesting that the use of manures 'so expensive that the farmer becomes blinded to their necessity' would be mitigated within the Garden City model, expressing 'the plan proposed embraces a system of sewage disposal which will return to the soil in a transmuted form many of those products the growth of which, by exhausting its natural fertility, demand elsewhere the application of manures' (Howard, 1902 [2014]). Howard had clearly taken inspiration from, and reinforced this argument using an extract from, Victor Hugo's, *Les Miserables* of 1862: 'When drainage everywhere, with its double functions, restoring what it takes away, is accomplished … the products of the earth will be increased ten-fold'. Howard takes great care to outline the metrics of his proposal. Using these metrics and comparing them to recent studies by, among others, Stockholm Environmental Institute, Cornell University and Matthew Wheeland (Pure Energies), we are able to test the validity of the Garden City model.

Testing the Garden City model

Taking the figures proposed by Howard for the overall area and population of a Garden City it is possible to quickly establish the land-share available per person.

Through an analysis of the current trends of consumption in the UK, it is possible to estimate the average land-share per person required. The validity of the Garden City as a model for urban sufficiency can then be tested through a comparison drawn between the productive capacity of the proposed Garden City model and requirements of a modern population. Throughout this chapter, land-share refers to the total area of land available or required per capita. A simple division of the total population of Howard's model by the total area of the Garden City shows a land-share of 0.07ha per capita. This initial figure reflects the total 2428ha (6000 acres) of Howard's Garden City model. Taking into consideration the zoning of Howard's proposal as previously discussed, a more accurate calculation would be to divide the 32,000 population by the 5000 acre (2023ha) agricultural belt; this equates to a land-share of 0.06ha per capita.

A 2011 study by Matthew Wheeland (2011) details the total hectares required to sustain a diet of 2300 calories per person per day. The study outlines the area required to grow sufficient fruit, vegetables and grains to meet this recommended daily calorie intake. The study goes on to outline the additional land required to support various methods of dairy and meat production. The inclusion of these metrics enables the calculation of total hectares required to sustain a wider parameter of diets inclusive of a vegetarian diet, white meat only (chicken and pork) diet and a diet that includes red meat (chicken, pork and beef).

Vegan diet

Wheeland suggests that in order to sustain a sufficient vegan diet of fruits, vegetables and grains per capita, per year, an area of 0.17ha is required (Wheeland, 2011). An initial comparison of the 0.17ha suggested by Wheeland to the 0.06ha per capita Howard's proposal can sustain immediately highlights an obvious shortcoming in the Garden City model. These figures suggest the Garden City would, in reality need to cover an area of 5440ha, almost three times that of Howard's proposal, if it were to sustain the population of 32,000 on a basic vegan diet.

Vegetarian diet

The impact an inclusion of diary products can have on land-share varies depending on the source. On average, 187.7 liters of cow's milk is consumed, in various forms, per capita, per year in the UK (Agriculture & Horticulture Development Board, 2014). To accommodate one dairy cow requires 0.5ha (Soil Association, 2009). A single dairy cow in the UK yields an average of 7327 liters of milk per year (Agriculture & Horticulture Development Board, 2014). Assuming the average of 188 liters is consumed per person, one cow could provide enough milk to satisfy 39 people. This equates to an additional 0.01ha required per person. On average, a single cow is fed 50kg of high-protein soya-based feed per day. In 2009 the European Union imported well over 45 million tons of protein-rich animal feed, consisting mainly of soybeans grown in South America, the growth of which requires roughly 1.2 million hectares of arable land. It can therefore be estimated that the additional land required to grow the feed for the dairy cattle requires an additional 0.14ha land-share per capita. However, if the residents of the Garden City were to source the milk for their dairy products from Nubian goats, as Wheeland suggests, two goats could provide enough milk for 18 people with only an additional 0.0002ha required per person (Wheeland, 2011).

In 2013, an average of 185 eggs were consumed per capita (British Lion, 2014). The laying capacity of hens varies with breed, ranging from approximately 170 eggs per year to 330 eggs per year (Happy Chicks, 2014). It is therefore estimated that one hen is required to meet the current per capita demand for eggs. The latest guidance by the Department for Environment, Food and Rural Affairs (DEFRA) recommends that, for the welfare of the hen and to be considered 'free range', each hen should have access to an area of $1m^2$ (CIWF, 2013). In addition to the land required to accommodate the livestock, additional crops will be required as feed. Wheeland suggests that an area of 0.006ha would provide enough corn to feed a single person and the associated livestock, although it is not clear whether this is total feed or a supplement to a grass-based diet. Advancing on the previously established requirement of 0.17ha (0.42 acres) to sustain a vegan diet, a vegetarian diet, presuming a substitution of cow's milk with goat's milk and including additional grain required as feed, would equate to a land-share of 0.18ha (0.44 acres) per capita. A vegetarian diet consisting of cow's milk would require land-share of 0.33ha per capita, almost six times larger than that proposed by Howard.

Diet including white meat (chicken and pork)

The impact introducing meat to a diet has on land-share is again relative to which meat is being farmed. In 2013, chicken was the most consumed meat in the UK, with 95 per cent of the population incorporating chicken into their diets at least twice a week (British Poultry Council, 2014). It is estimated that a total of 1270 million birds were consumed within the UK last year, and of this, 870 million were produced in the UK with another 400 million being imported (British Poultry Council, 2014). These figures suggest that an average of 21 chickens are consumed per person per year. Referring back to the welfare guidance of $1m^2$ per chicken (CIWF, 2013), the current trend of poultry consumption increases land-share per capita by 0.002ha. It is worth noting that these figures do not include the additional land requirements for reproduction or processing, which will obviously increase the land-share further and therefore this estimate reflects the best possible scenario. Wheeland (2011) suggests that three pigs would provide enough meat to sustain a family of four twice a week, and he goes onto suggest that each pig requires an area of $69ft^2$ (0.0006ha).

DEFRA guidelines, however, outline a minimum area of 0.04ha as required for a single sow. The Soil Association argues that a minimum area of 0.06ha per sow is required for organic certification (Soil Association, 2009) and 2.7kg of high-protein soya-based feed is the recommended quantity of feed per pig per day (Soil Association, 2009), which requires an additional 0.89ha per pig. If the diet within the Garden City were to be restricted to white meat only, and we were to consider the impact pig farming would have on the Howard's model, it could be estimated that this would increase the land-share to 0.46ha per capita assuming one pig per person, per year, farmed to DEFRA welfare standards. An analysis of these metrics show a minimum land-share of 0.46ha per capita is required in order to sustain a white meat diet. To provide the 32,000 population of Howard's Garden City with this land-share per capita would require a total agricultural belt of 14.720ha; over seven times larger than the initial proposal put forward by Howard. In fact it would require 7.3 of Howard's cities just to meet the food requirements of one Garden City. A per capita land-share of 0.46ha assumes a diet consisting of goat's milk as a substitute for cow's milk. If, however, cow's milk were to be introduced into the diet, the required per capita land-share would increase to 0.61ha and require an agricultural belt of 19,520 ha.

Diet including red meat (beef)

In 2012, 9.7 million cattle were farmed in the UK (DEFRA, 2012), accounting for 82 per cent of the total beef consumed (Global Food Security, 2014). This equates to approximately one cow being farmed a year per five people. Guidelines published by the Soil Association state that an area of 1.4ha is required to support ten beef cattle (Soil Association, 2011), plus the additional 5.48ha required to supply sufficient feed. This suggests an additional area of 1.4ha is required per person. Based on the figures discussed, sustaining a diet of fruit, vegetables, grains, dairy and meat to the current trend of consumption would require a total land-share of between 1.9ha and 2.01ha depending on the whether welfare or organic farming methods are followed. To provide a land-share of 2.01ha per capita would require an agricultural belt of 64,320ha. This is almost 32 times larger than that proposed by Howard and equates to a total of 31.8 agricultural belts, of Howard's design, being needed to sustain the city's 32,000 population.

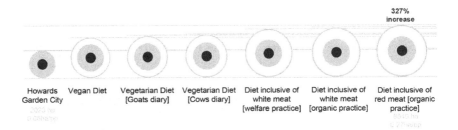

Figure 6.2 Howard's Garden City model shown in relative scale to estimated UK land-share of various diets

Reflecting on Howard's Garden City model

The Garden City model, as designed by Howard would appear to satisfy the claims of a model that combines the positive aspects of urban with those of rural. Wilson's biophilla hypothesis (1995) illustrates the importance of such an ecological model; he argues that humans have an innate attraction to life and life's processes (Wilson, 1995). Howard's Garden City model could also be considered to be 'one of the first manifestations of sustainable development' (Town and Country Planning Association, 2014) as it does include many basic but key principles of sustainable design such as localized food production (Newman and Jennings, 2008), energy production (Van den Dobbelsteen, 2010) and nutrient recycling (Girardet, 2011). However, the comparison of available land-share to required land-share highlights the extent to which the Garden City model, contrary to popular description, is inept at self-sufficiency, at least in terms of food production.

The current design of the Garden City model is incapable of sustaining a population of 32,000 people, as Howard suggests. The analysis of current UK consumption and land-share requirements indicates that a model that consists of 2023ha of productive land is only capable of producing 3.1 per cent of the food required to sustain such a population with a diet of fruit vegetables, grain, meat and dairy. Alternatively, on the same diet, the Garden City model as proposed could sustain a population of 1006 people, or 11,900 people on a vegan diet.

Newman and Jennings (2008) argue that cities work best at bioregional and community scales, not at the global scale at which they are currently performing. Newman and Jennings suggest the importance of urban consumption and production to be better matched to bioregional capabilities. It could therefore be argued that if the Garden City is to be considered as a model of ecological urbanism, it must be updated to reflect the land-share requirements of its population.

Updating the Garden City model

Jane Jacobs suggests that cities are an excellent opportunity to initiate change. She argues cities are crucibles for innovation and the source for global change, suggesting there are many examples throughout various stages of history where innovations from one city have been rapidly copied and improved by others (1961, 1984; Newman and Jennings, 2008). 'The best human innovations mimic and learn from natural systems and cities need to reflect this approach to innovation in their planning, design, production and consumption' (Newman and Jennings, 2008). Herbert Girardet (1999) describes a sustainable city as, 'organized so to enable all its citizens to meet their own needs and to enhance their well-being without damaging the natural world or endangering the living conditions of other people'.

Many interpretations of sustainability generally centre on creating and maintaining the conditions under which humans and nature can exist (Ferrao and Fernandez, 2013). Rob Hopkins (2008) would argue that sustainability is not enough, and what designers should be aiming for is resilience: the ability to absorb disturbance and reorganize while undergoing change, so to retain essentially the same functions. To achieve this, it could be argued that we need to consider Jane Birkeland's proposal of positive development. Birkeland suggests that the built environment can create the infrastructure, conditions and space for nature to continue its life-support services and self-maintenance functions by approaching the planning, design and maintenance of the urban environment as an opportunity to expand the ecological base (Birkeland, 2008).

Nested redundancy

The theory of 'nested redundancy' (Keeffe, 2014) argues that systems and processes should be designed to be closed loop in all scales of architecture and urbanism from the individual through to individual dwellings, streets and neighborhoods, and then the city and finally the bioregion. The current practice based on a modernist reading of the city of concentrating productive infrastructure at a certain scale, and having none at any other scales, is flawed as it creates a system reliance on only one source, which does not promote resilience. Nested redundancy aims to scatter productive infrastructures at a range of scales, linked with smart grid technology (or the equivalent), so that in the event of a catastrophe, which might wipe out production at one scale, there will still be some output available from other sources that allow the city to continue to function (albeit at a lower subsistence output). An example would be the embedding of energy infrastructures, where the roof of a house might provide top up electricity, whilst a street-scale combined heat and power plant could provide a more general base load of heat and power, whilst a regional wind farm or biomass plant might provide the peak load requirement. Thus if one scale failed, there would be a reasonable supply of resource still available. The system can be applied across energy, food and other resource streams, and helps to embed resilience within the city. Marcotullio and Boyle (2003) also support the view that sustainability can only be achieved when cities are approached as systems and components of nested systems in ecological balance with each other.

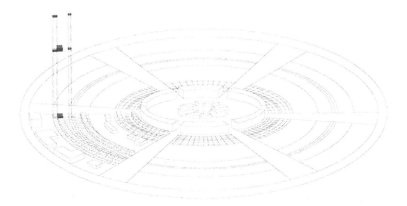

Figure 6.3 Nested redundancy: dwelling scale

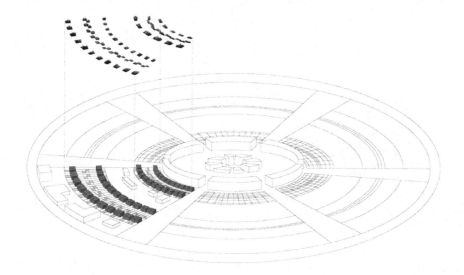

Figure 6.4 Nested redundancy: street scale

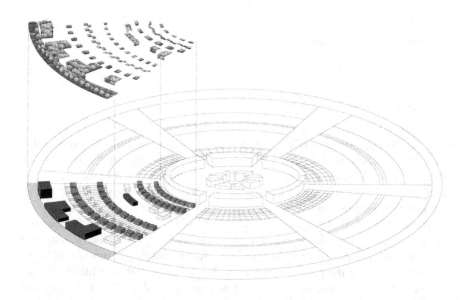

Figure 6.5 Nested redundancy: neighbourhood scale

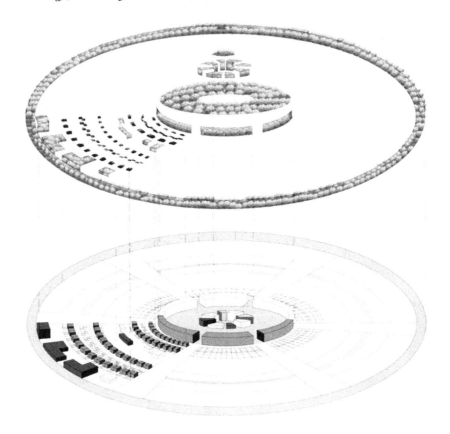

Figure 6.6 Nested redundancy: city scale

Integrated urban food production

The principles of the Garden Cities in many respects do attempt to address the issues of the sustainability of cities, and it could be suggested that as a model that promotes localized food production, it could potentially improve the resilience of the UK as a whole due to dramatic decrease in the reliance on food imports. It could however be argued that Howard's model, through its significant use of zoning does little to increase local resilience and limits the ecological base.

An alternative to Howard's zoned approach could see a number of productive pockets created throughout the new Garden City model. Creating a system of multiple growing platforms within the city could increase biodiversity within the area and minimize the potential impact of disease, among other positives, increasing the cities resistance to shock and improving local resilience. Ferrao and Fernandez (2013) suggest that cities are 'certainly not sites for agricultural production'. Carolyn Steel in *Hungry City* opposes this and suggests that cities have often been used as agricultural hubs in the past and should be again. During the Dig for Victory campaign during the First World War it is estimated that allotments were providing half of all Britain's fruit and vegetables (Steel, 2008).

Returning back to the original analysis of Howard's design, he described a number of 'open spaces'. In addition to increasing resilience and expanding the ecological base, the integration of food production into the heart of the urban model increases the productive capacity of the city. The 'Garden' located in the centre of Howard's model is described as 'five and a half acre beautiful and well-watered garden'. If, however, this land was used in a productive capacity, it could potentially provide the Garden City with an additional 2.2ha of productive landscape. 'Grand Avenue', which Howard describes as 'an additional park of 115-acres', could potentially provide an additional 46.5ha worth of growing area. Finally, 'Central Park' described as '145 acres wide, extending in a circle of over three miles' is yet another potential area in which to incorporate urban food production. If utilized, Central Park could present the Garden City with a further 58.6ha of productive land.

Crystal Palace encircles Central Park and was designed by Howard with a primary function of an indoor market place; however it was designed 'a good deal' larger than required to enable a considerable part of it to include a Winter Garden. Using the metrics outlined by Howard it can be estimated that the circumference of Crystal Palace would be 2815.6 meters in length and approximately 16 meters in width. On the assumption that half the total area of the Crystal Palace was utilized for the purpose of an indoor market and the remaining half be divided between a Winter Garden and an indoor growing area, an additional 1.08ha could be used in a productive capacity.

Within his proposal, Howard suggested an average plot size of 20ft x 130ft; this equates to approximately 0.02ha per plot. Currently the average plot size in the UK measures 335m^2, equating to 0.03ha per plot. If the average plot size within the new Garden City model was reduced from the current UK average to reflect Howard's proposal, each plot could include a 0.01ha 'garden growing' area. Increasing the productive area of the Garden City by 0.01ha per plot, assuming Howard's estimation of 5455 individual plots, would equate to an additional 54.5ha of productive land. Potentially, each plot, if only using the minimum allotted area for food production, could produce a further 2000 salad crops.

Layering technologies

Howard hinted at the role technologies could play in future urban models, suggesting that all machinery within the Garden City would be 'driven by electric energy'. There have been huge advances in technology since Howard first designed the Garden City, and just as

Howard embraced electric powered machinery, new technological advances should be incorporated into a new urban model. Various technologies can be implemented at all urban scales, from smart houses that ensure maximum energy efficiency, to bio-productive facades on high-rise office buildings, to solar farms

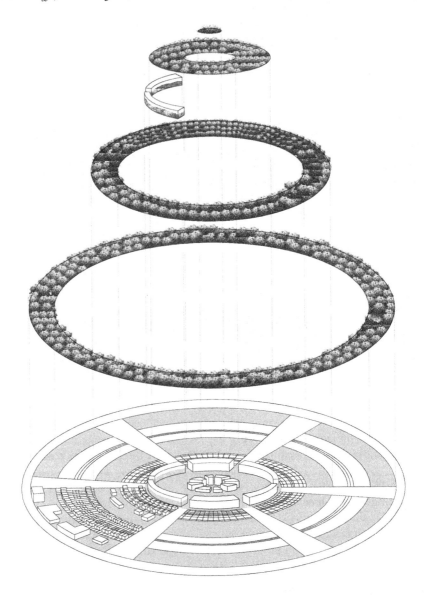

Figure 6.7 Integrating food production increases the city's productive capacity while also increasing the ecological base and local resilience

along the city perimeter. DEFRA has estimated the total available agricultural land in the UK to be 18,349,000ha. Assuming a total UK population of 64.1 million (ONS, 2014), this equates to a per capita land-share of 0.28ha. This suggests that based on the current consumption rate, the UK has already reached its productive capacity. Incorporating new technologies can allow previously unproductive areas to be used for food production. In addition to the current two-dimensional agricultural model that sees productivity restricted to the horizontal plane, new technologies, such as those demonstrated by Jenkins *et al.* (2014) can release vertical and elevated surfaces to be used for food production, advancing to a three-dimensional urban model.

The Crystal Palace for example, proposed by Howard as 'a wide glass arcade' was intended to be 'one of the favorite resorts of the people'. The circumference of the Crystal Palace has previously been estimated at 2815.6m in length. Assuming both a height of 22m, matching that of the Winter Gardens in Sheffield, and productivity restricted to the south-facing portion of the facade, the Crystal Palace has a potential vertical growing area of 30,971.6m2, an additional 3.09ha.

Recent research collaborations between by Queens University Belfast, Siemens, BDP and Glass Solutions produced a prototype of a modular food producing facade. Powered by photovoltaic panels and using an integrated aquaponic system, the facade combined small-scale aquaculture with soilless growing techniques to vertically grow a variety of salad crops and edible fish. The working prototype revealed that using this system it was possible to grow an average of 15 salad crops per meter square of facade, taking into consideration the necessary structure and structural openings (Jenkins *et al.*, 2014). The development of this new technology suggests that, should this facade system be incorporated onto the south facade of the Garden City's Crystal Palace, it could potentially produce an additional 464,574 crops. Assuming a harvest every eight to nine weeks for a limited growing season of eight months a year (Jenkins *et al.*, 2014), this productive facade could provide the Garden City with an additional 1,858,296 crops per year.

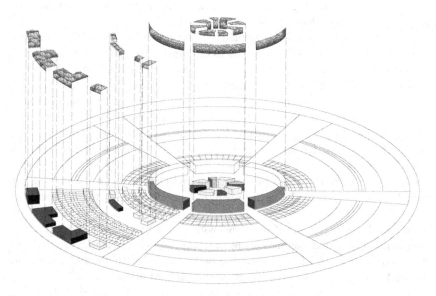

Figure 6.8 Incorporating new technologies allows previously unproductive areas to be used for food production

Conclusion

Following a detailed analysis of Howard's Garden City proposal it would appear the model could be considered to be an early example of sustainable urban design. It responds to the issues resulting from rapid urbanization with a number of key principles associated with ecological urbanism, such local food production and nutrient recycling.

A detailed analysis of the metrics provided by Howard and those of the current consumption figures within the UK enabled a comparison of the two sets of data, meaning it was possible to test the validity of Howard's proposal of the Garden City model as being one of a sustainable solution to urbanization.

The metric analysis shows that for a population of 32,000, as proposed by Howard, each person, sustaining a strict vegan diet, would require a land-share of 0.17ha. To provide this, based on Howard's model, an agricultural area of 5440ha would be required. This is a 168.9 per cent increase on the 2023ha proposed by the Garden City model. If the Garden City model were to supply a diet inclusive of meat, a land-share of 0.33ha would be required per capita for white meat (chicken and pork), only resulting in an agricultural belt of 10,560ha. To also supply beef, a land-share of 2.01ha would be required per capita and a total agricultural belt of 64,320ha. This represents a 2180 per cent increase, which would require an additional 31.8 agricultural belts as designed by Howard.

As a model for ecological urbanism, the Garden City model is flawed, incapable of supplying more than 3.1 per cent of the food required by the city's population. The use of zoning is highly inefficient in terms of productivity and in establishing an ecological base; nor does it add any layers of resilience to the model, as such, it can be argued that this highly regarded urban model is not a valid solution to the effects of increasing urbanization.

An analysis of the data has presented a better understanding of the land-share requirements within the UK, as well as an opportunity to update the Garden City model to a model that better reflects urban consumption and aims to match it through urban production. In contradiction to the zoned model of Howard's Garden City, integrating urban food production to establish a network of productive pockets increases resilience and the ecological base of the city. Utilizing urban parks and open spaces as productive areas can increase the productive capacity of the Garden City model. Layering new technologies can increase efficiency and release previously untapped productive potential. Incorporating technological advancements such as the bio-productive facades can transform the Garden City model from its previously two-dimensional model into a three-dimensional productive urban landscape.

Approaching the design of our cities with nested redundancy at the forefront, integrating urban food production and layering technologies allows us to develop the Garden City model into what Howard intended it to be, and what many believed it to be: a model for self-sufficient ecological urbanism.

References

Agriculture and Horticulture Development Board (2014) UK, GB and NI Farmgate prices. http://dairy.ahdb.org.uk/market-information/milk-prices-contracts/farmgate-prices/uk,-gb-and-ni-farmgate-prices/.

Barrett J. (2003) *An Ecological Footprint of the UK: Providing a Tool to Measure the Sustainability of Local Authorities*. Stockholm Environment Institute, York.

Birkeland, J. (2008) *Positive Development: From vicious circles to virtuous cycles through built environment design*. Earthscan Routledge, Oxford.

British Lion (2014) Egg info. www.egginfo.co.uk/egg-facts-and-figures/industry-information/data.

British Poultry Council (2014) British Poultry Council. www.britishpoultry.org.uk.

CIWF (2013) *Animal Welfare Aspects of Good Agricultural Practice: Chicken Production. Broiler Chicken Case study UK 2: RSPCA Freedom Food Free-range farm*. RSPCA, Horsham.

Cooper, N. and Dumpleton, S. (2013) *Walking the Breadline: The scandal of food poverty in the 21st century*. Oxfam, Oxford.

DEFRA (2012) *Farming Statistics: Final Crop Areas, Yields, Livestock Populations and Agricultural Workforce at 1st June 2012*. DEFRA, London.

Ferrao, P. and Fernandez, J. (2013) *Sustainable Urban Metabolism*. MIT Press, London.

Girardet, H. (1999) *Creating Sustainable Cities (Schumacher Briefings)*. Green Books, Devon.

Girardet, H. (2011) *Creating Sustainable Cities: Schumacher Briefings 2*, Green Books, Devon.

Global Food Security (2014) www.foodsecurity.ac.uk/food/food-global.html

Happy Chicks (2014) Happy Chicks. www.happychicks.co.uk/speckeldy-hens-173-p.asp.

Hopkins, R. (2008) *The Transition Handbook*. Chelsea Green Publishing, Vermont.

Howard, E (1902 [2014]) *Garden Cities of To-Morrow*, Kindle Edition.

Jacobs, J. (1961) *The Death and Life of Great American Cities*. Vintage Press USA.

Jacobs, J. (1984) *Cities and the Wealth of Nations: Principles of Economic Life*. Random House USA.

Jacobs, J. (2001) *The Nature of Economies*. Vintage Books, London.

Jenkins, A., Keeffe G. P. and Hall N. A. (2014) Façade Farm: Solar mediation through food production. In Frank and Papillon (eds) *Proceedings of the ISES EuroSun 2014 Conference*. ISES Freiburg.

Keeffe, G. P (2014) *Super Suburbia: Developing resilience by closing resource cycles in North Belfast*. USAR, London.

Marcotullio, P.J. and Boyle, G. (2003) *Defining an Ecosystem Approach to Urban Management and Policy Development*. Tokyo: United Nations University Institute of Advanced Studies.

Newman, P. and Jennings, I. (2008) *Cities as Sustainable Ecosystems*, Island Press, Washington DC.

ONS (Office for National Statistics) (2014) *Measuring National Well-being: Life in the UK*, ONS, London.

Soil Association (2009) Organic Pig Production: An introductory guide. Bristol. www.soilassociation.org/LinkClick.aspx?fileticket=3zrS-i2tpWI%3D&tabid=133.

Soil Association (2011) Organic Beef and Dairy Production. Bristol. www.soilassociation.org/LinkClick.aspx?fileticket=EWCxnzMh0N0%3d&tabid=131.

Steel, C. (2008) *Hungry City: How food shapes our lives*. Vintage Books, London.

Town and Country Planning Association (2014) *The TCPA Garden City Principles Notes*. TCPA, London.

Wheeland M. (2011) How Big a Backyard Would You Need to Live Off the Land? https://pureenergies.com/us/blog/live-off-the-land-2/.

Van den Dobbelsteen, A. (2010) Use your Potential! Sustainability through local opportunities. Inaugural lecture, TU Delft June.

Wilson, E. (1995) *The Biophilia Hypothesis*. Island Press, Washington DC.

World Health Organization (2014) www.who.int/research/en/.

7 Urban agriculture up-scaled

Economically and socially productive public green space

Joshua Zeunert

Introduction

'Space is fundamental in any form of communal life; space is fundamental in any exercise of power' (Foucault & Rabinow, 1984: 252). Public green space has the potential to provide one our last remaining free sources of access to open land, clean air, vegetation, water and soil within the urban realm. In most developed countries, this space – due to complex, interconnected legacies of enclosure, privatisation, population growth, urbanisation and 'modernisation' – typically exists as controlled, contrived, scenic picturesque landscapes, unavailable for forms of civic, productive and generative activities at scale, such as public urban agriculture. Narrow assessment of green space's on-going financial and maintenance costs fail to recognise wider gains (such as physical and psychological wellness, increased property values, decreased crime rates) (Maller, 2002; Woolley, 2004; Sherer, 2006) and despite attempts, studies that present financial benefits of green spaces have not yet managed to stem the tide of budget cuts and reduced spending. Perhaps more importantly, income-generating strategies within public green spaces have not been sufficiently explored. Such approaches could help to develop more convincing arguments analogous with the measurement metrics and quantitative language threatening green space's optimisation and survival. By 'up-scaling' public green space's productive capacity within an ethical framework, we have the potential to greatly enhance social and environmental performance – shifting the existing paradigm from passive to active, consumptive to generative and centralised to collective.

Reduced spending and quality of green space

Urban public green spaces (defined as parks and landscaped spaces under public ownership within urban areas but excluding peri-urban green belts and conservation areas) usually require external, off-site sourced budgets to meet financial costs for their maintenance and upkeep. In 'income versus expenditure' terms, most projects or spaces cost municipal authorities more money than they generate. The Adelaide Parklands, for example, annually costs the Adelaide City Council AU$9 million for maintenance ($11,842 per hectare (/ha) and $22,895/ha maintenance and capital works), yet the annual income generated is below $1 million (Cook, pers. com., 2014), while the (British) National Audit Office estimates that the cost of maintaining and renovating urban green space (in England) was UK£700 million in 2004/05 (NAO, 2006). In the Western world's present politically austere climate, a central focus on the financial bottom-line routinely affects public services, whereby existing public green spaces are frequently targeted for cost-cutting, leading to reduced expenditure and upkeep (Dunnett *et al*, 2002; CABE Space, 2004: 3; Loukaitou-Sideris, 2006; Sherer, 2006; Foderaro, 2014).

Budgets for green space expenditure are heavily reliant on the often-unstable fiscal and planning cycles of politics and governments (NAO, 2006; Sherer, 2006: 12). In recent decades in the USA, UK and Australia, the overall trend for maintenance spending on existing green spaces has been in decline (NAO, 2006) – heightened since the 2008 global financial crisis (local authority spending on open spaces in the UK was, for instance, cut by 10.5 per cent between 2010/11 and 2012/13 (Drayson, 2014: 8)). Reduced budgets in turn lead to degradation and lower appeal (Sherer, 2006), often resulting in undesirable and unsociable activities (CABE Space, 2004: 4; Drayson, 2014: 7). Vandalism, graffiti and illegal dumping cost municipal authorities additional revenue (SPUR, 2012) as well as increased output for surveillance, lighting and deterrents of anti-social behaviour.

In order to be successful, the introduction of economically generative land uses need to both meet public expectations and to operate within the common good (Linn, 1999). Traditional measures of economic viability (revenues, jobs, profits) must be used, if for no other reason than to attract the attention and support of funders and policy-makers (Smit in Lazarus, 2005: 22).

Securing a reliable income source and increasing sociable usage are, however, in the interests of both participants and green space designers and managers (planners, landscape architects, municipal authorities, open space managers and maintenance staff) (Sherer, 2006). This chapter explores the gross economic potential that urban agriculture could contribute to existing green spaces within developed countries (although calculations are not intended to be directly applicable to any specific site). Gross financial sums are calculated and provided incrementally, relative to instigating urban agriculture as a percentage of the area of several existing green spaces (10–50 per cent). This research does not intend to examine or propose appropriate funding, regulatory, legal and planning models or structures for public urban agriculture in green space as this requires further research and trialling (Cabannes & Dubbeling, 2001). Instead, the chapter tests scenarios using simplified modelling, raises issues that require further attention and makes recommendations based on findings.

Green space themes and usage

Internationally, there have been shifting, contested, lobbied and debated themes of urban green space land use (Loukaitou-Sideris, 2006) The mid to late twentith century has focused on visual amenity, passive leisure, recreation, sport, respite and tourism (Corner, 1992), within a Western cultural historical aesthetic framework of the 'picturesque', 'countryside', 'aristocratic' and 'estate'. Accompanying behavioural expectations tend to the passive rather than active (with the exception of sporting activities), constrained by a range of control mechanisms such as fencing, signage ('keep off the grass', 'no ball games') and activities such as walking, sitting, promenading and viewing. Direct participation or the act of 'getting one's hands dirty' is uncommon – participants typically look but do not touch – more subjugated than participatory sharers and shapers of land (Linn, 1999).

There have been numerous attempts to overcome contrived, visual notions of landscape in the past two decades through restoration ecology (Leeuwen *et al*, 2010: 20–21), predominantly through a nativist plant ethic (Zeunert, 2013). Despite benefits of more sustainable landscaping (such as increased biodiversity, less water usage, suitability to local context), results are ultimately similarly passive: fledgling ecologies are precious and to be protected, viewed and appreciated, not engaged with in a hands-on manner. Additionally, dense urban populations and highly trafficked environments (by humans, domestic and wild predators) are typically counter-productive to fragile biodiversity, thus restoration ecology is often not an optimal urban land-use theme (Zeunert, 2013).

> We are slaves in the sense that we depend for our daily survival upon an expand-or-expire agro-industrial empire—a crackpot machine—that the specialists cannot comprehend and the managers cannot manage. Which is, furthermore, devouring world resources at an exponential rate.
>
> (Abbey in Berry, 1985: 38)

Agriculture has featured prominently in urban green spaces throughout civilised history (Leeuwen *et al*, 2010), but is currently an uncommon, marginalised or minute land use (Grewal & Grewal, 2012). It rarely features as a permanent theme or substantial land area in prominent urban green spaces.

In contrast to passive land use, agriculture, especially urban agriculture, involves direct, active and on-going participation and activation. Alon-Mozes (2009) posits that the tangibility of food production is of vital importance to an increasingly urban, global and homogenous population, with local food production playing a key a role in regional identity, cultural heritage, distinctiveness and competitive advantage (Mason & Knowd, 2010). Concerns about the on-going viability of our food supply are increasingly becoming the subject of public and policy discourses under the phrase 'food security', strengthening the argument for increased implementation of urban food growing.

Crisis

During the twentieth century, food production in public green spaces has been deployed in times of crisis. In many cases it has provided significant yields, for example, during World War II, 40 per cent of vegetable consumption in North America (Brown & Jameton, 2000), and around half of Britain's fruit and vegetables (Garnett, 1996). Public food production assisted economic meltdown in Rosario, Argentina, in 1989 and 2001, provided over 40 per cent of vegetables and small livestock during blockades in Sarejevo in 1992 (Sommers & Smit, 1994), and continues to provide around 90 per cent of vegetables in Havana, Cuba, long after the Soviet collapse in 1989. Urban agriculture, however, can and should offer benefits well beyond crises stemming from political mismanagement. Urban farmers in Singapore and Hong Kong produce 30–50 per cent of fresh produce in these densely urbanised, economically advanced and high-income cities, and 76 per cent and 81 per cent of vegetables in China's Shanghai and Beijing (Sommers & Smit, 1994; Halweil, 2004). Poorer cities often have high urban agriculture outputs. Hanoi, Vietnam, for example, grows 80 per cent of its fresh vegetables in an urban context (Sommers & Smit, 1994), and Dar es Saalam, Tanzania, over 60 per cent of milk and 90 per cent of leafy vegetables (Lee-Smith & Prain, 2006). After economic recovery or following economic development, public space urban agriculture commonly returns to landscaping with ornamental plant growth or shifts to built development (such as housing). This transition can stem from concerns regarding urban agriculture and public health, but more often, urban agriculture's erasure reflects a misguided desire for 'urban modernity' (Morgan, 2013: 20).

Urban agriculture literature

Most urban agriculture literature in developed countries examines social (Patel, 1991; Malakoff, 1995; Garnett, 1996; Pretty, 2002; Flores, 2006; Mason, 2006; Sumner et al, 2010), environmental (Halweil, 2004; Doron, 2005; Flores, 2006; Merson et al, 2010: 79) and physical, psychological and health (Kaplan, 1973; Brown & Jameton, 2000; Hassink et al, 2007; Duchemin et al, 2008; Blaine et al, 2010) issues and benefits. As noted by McClintock et al, it is challenging to capture urban agriculture's multiple, beneficial attributes (2013: 55), yet when appropriately executed, the practice can be complementary to 'notable' green space attributes being 'esthetic, recreational, educational and health benefits' (McClintock et al, 2013: 55) and biodiversity (Garnett, 1996). The small amount of economic literature (Moustier & Danso, 2006; Masi, 2008) reflects the phenomenon that urban agriculture usually exists at a community or local scale (Pudup, 2008). It is telling to note that Golden's literature review of urban agriculture determined that economic literature was 'very limited' (2013: 13).

Scale of urban agriculture

The allotments, community gardens, urban farms and private enterprises constituting urban agriculture are generally of small scale, ranging between 0.01 and 3ha, and therefore generating correspondingly low yield and economic volumes. The 250,000 community and allotment gardens in the Netherlands, for example, total around 4,000ha (Leeuwen et al, 2010: 22) with an average of 0.016ha (160m²) each. The peri-urban fringe is often the location of larger agricultural activities due to lower land values than inner urban areas. Here, significant scope exists for up scaled social and public enterprises, for example, the City of Enfield Council in the UK is pursuing a market garden strategy aimed at generating 1,200 jobs catalysed through urban agriculture interventions on municipal land (Enfield, n.d.).

Inner city urban agriculture is usually 'grass roots', community-led or small entrepreneurial growing interventions rather than highly organised or professional growing operations (Cities, 2013). Yields may not be the primary focus for individuals or groups because their objectives are informed by social goals (Nettle, 2010: 1; Firth *et al*, 2011: 557) or recreational reasons (Leeuwen *et al*, 2010: 21). In this social context, the growing costs often exceed relative economic value of yields (Leeuwen *et al*, 2010: 22).

Most community and allotment gardens have zoning or by-laws stipulating growing only for individual and non-commercial purposes. Commercially viable, small-scale, private urban agriculture interventions are, however, increasingly appearing, especially in North America (Abelman, 2005), but most are reliant upon grants, donations, crowd-sourced funding or entrepreneurial measures to establish and sustain their on-going viability. Adding to challenges faced within the urban context is the fact that much urban agriculture is temporary or transitory, occurring on vacant land and with 'meanwhile' leases. These interventions can stimulate local economic development but also face displacement at any point in time. While some practitioners (such as Chicago urban agriculturalist Ken Dunn) see this as a measure of success (Finkel, 2012), the operators of Dalston East Curve Garden in London, for example, find this prospect frustrating, having establishing a successful local enterprise (LI, 2013). Facilitating permanent urban agriculture in existing public green spaces could mitigate displacement issues, attract increased investment and provide much needed stability for growers.

Methodology

Due to limited economic examination (Golden, 2013: 13), no known directly related literature is currently available to provide comparison. Ranging methods are examined to assess the economics of production approaches to compare possible $/ha returns. Several well-known green spaces of various sizes in North American, Australian and European cities have been presented because they provide potential familiarity with their scale and possess proximity to urban populations of half a million or more (Table 7.1). While this would also ensure high local consumption of fresh produce as well as workforce availability, their inclusion is not intended to suggest their suitability to urban agriculture, which would be determined through detailed, site-specific studies. The methodology employed explores gross annual economic returns based on calculations of horticultural crop yields and economic data from eight sources. Fourteen US$/ha figures have been calculated based on differing growing approaches (Table 7.2). These $/ha figures are multiplied at incremental scales from 10 to 2,500 hectares, with public urban green spaces in Table 7.1 based on < 25 per cent and 25–50 per cent of their total area used under each of the 14 $/ha hectare growing scenarios (Table 7.3). The production methods examined are as follows.

Organic wholesale USA

Organic vegetable and herb growing in the USA is calculated through 2011 national survey data (USDA, 2012) and determined by dividing the total harvested cropland area by the total gross value of sales (data converted to metric).

Conventional retail UK

Yield data from the highly distributed Royal Horticultural Society's (RHS) *Fruit and Vegetable Gardening* (Pollock, 2002) was converted to hectare values using their recommendations for plant and row spacing. If a yield range was provided by the RHS, an average is calculated. Spacing between rows (for trees), which was not provided, is calculated at 1.5 times plant spacing. A 96x96 metres growing area is used/ha; 12.75 per cent/ha is excluded for infrastructure and other necessary non-growing areas. Yield data is multiplied by average retail price data from the UK Office for National Statistics (2013), although limited retail data were provided for fresh produce. Accordingly, values for three commonly available items are calculated (onions, carrots and apples). Although not shown, calculations for pears and grapes produced similar gross values/ha. Calculations are based on single cropping as discussed in the results.

Organic farmers market Australia

Calculations based on The Food Forest organic 15ha permaculture property in Australia (Brookman, pers. com., 2011) use three basic crops and their yield/ha. Economic data reflect the Food Forest's approach of selling direct at farmers markets in Adelaide, Australia.

Diversified urban farm ideal climate USA

Fairview Gardens is an organic 4.8ha non-profit and educational farm in Goleta, California, which in 2005 grossed over $700,000 (Ableman, 2005). A calculation is achieved by dividing the economic return by the area (data converted to metric).

Specialty crops USA

Specialty crops are those grown with the intention of returning a high economic return/ha. *Profitable Plants Digest* is a USA-based publication focusing on growing profit-oriented plants, which are often specialised and tailored for local demands. Current examples include lavender, gourmet mushrooms, woody ornamentals, landscaping trees and shrubs, bonsai plants, Japanese maples, willows, garlic, bamboo and herbs (Wallin, n.d.). Data cited by editor Craig Wallin state that $USD60,000 per acre ($US148,263/ha) is 'not unusual' (Wallin, n.d.) and up to $90,000 per acre is possible ($222,394/ha); $148,263/ha was used in calculations (data converted to metric).

Urban farm limited growing season USA and Canada

Jean-Martin Fortier and Maude-Hélène Desroches cultivate 6,000m² of organic land, 'Les Jardins de la Grelinette' near Saint-Armand in Quebec, Canada, largely based on Eliot Coleman's techniques (below). Their annual sales are more than CA$100,000 (Martin, 2013: 35).

Eliot Coleman and Barbara Damrosch intensively farm vegetables on 5,600m² (>1.5 acres) at Four Season Farm in Maine in northeast USA, where 840m² is under greenhouses (one third of these are heated in winter). Around 55 vegetables are grown and sold, either freshly harvested or from storage 12 months of the year. Gross income from vegetables was $165,000 in the 2013 fiscal year (Coleman, pers. com., 2014). A calculation is obtained by dividing the return by the area (data converted to metric).

Intensive small plot urban farming USA

Wally Satzewich, Gail Vandersteen and Roxanne Christensen coined the term 'SPIN Farming', a small plot intensive approach. They provide metrics for calculating economic return using two measurements of 'low' and 'high' intensity (SPIN, n.d.). These were converted into $/ha values.

Results and discussion

Table 7.1 shows the $/ha annual gross return for each production method (lowest to highest). The results vary significantly, ranging from $22,249 to $360,000/ha.

The lowest values achieved are from an average of organic growers in the USA for vegetables ($22,249/ha) from a 2011 USDA production survey. A huge 88 per cent of USA's organic growers sold their produce to wholesale markets in 2008

Table 7.1 List of existing green spaces

Existing green space	Location	Total area (ha)	Total area (acres)	25% of area (ha)	50% of area (ha)
Washington Park	Chicago, USA	151	372	38	75
Royal Park	Melbourne, Australia	181	447	45	91
Central Park	Manhattan, USA	340	840	85	170
Hyde Park and Kensington Gardens	London, UK	253	630	63	127
Bushy Park	London, UK	444	1097	111	222
Englischer Garten	Munich, Germany	567	1400	142	283
Phoenix Park	Dublin, Ireland	707	1750	177	354
Adelaide Parklands	Adelaide, Australia	760	1878	190	380
Western Sydney Parklands	Sydney, Australia	5281	13050	1320	2641

(USDA, 2008: 4). Direct-to-retail and direct-to-consumer crop sales significantly increase economic return, although these approaches are usually inconvenient for rural farmers who prefer selling their larger volumes to wholesale buyers. Although no location data is provided in the USDA 2011 survey, the great majority of organic vegetable growers in the USA, averaging 24/ha of production (USDA, 2012: 18), are located in non-urban contexts where lower land values decrease the incentive to farm intensively.

The second lowest data resulted from onion production based on RHS yield data sold at UK retail prices ($35,164/ha), however, RHS yield data are based on single cropping. If multiple crops were grown in a year there would be a significant increase in economic return. The RHS data for carrots returned $46,685 per crop/ha, with the RHS stating that carrots are ready for harvesting 12–16 weeks after sowing. The growing season in central England is 275 days (DECC, 2013), enabling two to three harvests of carrots per year (without resorting to greenhouses) and thus returning $46,685/$93,370/$140,055 for single/double/triple annual cropping. Higher prices for fresh, locally grown produce might be obtained in comparison to the average retail price used for RHS calculations depending on the 'primary sales' method used (Table 7.2).

The Food Forest calculations vary, ranging from $38,000/ha for apple orchards (excluding additional income from non-saleable apples into juice, vinegar and cider production), to $99,750/ha for citrus orchards and $285,000/ha for the intensive annual production of organic lettuce. The $285,000 figure is a projection – yield and sales data indicate that it is possible though most likely unfeasible for non-wholesale distribution, thus lowering the return/ha (farmers markets will return a higher value for growers but there is a limited volume of one crop that can be sold in this context). Higher value crops (such as lettuce and herbs) are also more labour and nutrient intensive (than apples and citrus). Brown calculated a similar figure for lettuce production in Brisbane, Australia, with a result of $235,125 per ha (2008: 37).

RHS and UK retail calculations for apples returned similar prices to organic USA herb (wholesale) production ($41,293 and $41,645). Unlike multiple harvest annual crops, apples (and nuts) will yield a single harvest per year and are therefore likely to return lower values. Multiple harvests will increase gross returns but require greater labour.

Fairview Gardens generates an income of $145,833/ha through both direct marketing of produce (community supported agriculture, farm stand, farmers market and wholesale distribution) and a diverse range of activities (social programmes, events, training courses, volunteers, apprenticeships and donations). The economic return is an indication of gross income/ha for an innovative urban agriculture operation not solely limited to growing and selling produce. From the eight growing scenarios examined, Fairview Gardens is an ideal precedent to further explore and test public urban agriculture in green space due to its range of social and growing activities.

Craig Wallin's (n.d.) figures quoted in *Profitable Plants Digest* are comparable (though unsubstantiated through primary evidence). The *Digest* advocates growing niche products ideally requiring less intensive maintenance and supplying local

Table 7.2 Return in US$ per hectare for each production method

Growing system (annual gross return)	Location	Source	Conversion US$	Primary sales	Context	Per hectare
USA All Vegetables (Organic)	USA – average	(USDA 2011)		Wholesale	Rural	$22,249
UK Yield & Retail Data: Onion (Conventional)	UK	(Pollock, 2002; ONS, 2013)	(UK£– US$ 1.67)	Retail	Peri-urban	$35,164#
AUST Farmers market: Apples (Organic)	Gawler, Australia	(Brookman, 2011)★	(AU$– US$ 0.95)	Farmers market	Peri-urban	$38,000
UK Yield & Retail Data: Apples (Conventional)	UK	(Pollock, 2002; ONS, 2013)	(UK£– US$ 1.67)	Retail	Peri-urban	$41,293
USA Herbs (Organic)	USA – average	(USDA, 2012)		Wholesale	Rural	$41,645
UK Yield & Retail Data: Carrot (Conventional)	UK ONS, 2013)	(Pollock, 2002;	(UK£– US$ 1.67)	Retail	Peri-urban	$46,685#
AUST Farmers market: Citrus (Organic)	Gawler, Australia	(Brookman, 2011)★	(AU$– US$ 0.95)	Farmers market	Peri-urban	$99,750
SPIN low value	Philadelphia, USA	(SPIN, n.d.)		Direct marketing	Urban	$120,000
Fairview Gardens (Organic)	Goleta, California	(Ableman, 2005)		Mixed	Urban	$145,833
Specialty crops	USA	(Wallin, n.d.)		Direct marketing	Peri-urban	$148,266
Les Jardins de la Grelinette	Quebec Canada	(Martin, 2013)	(CA$– US$ 0.92)	Direct marketing	Rural	$151,561
AUST Farmers market: Lettuces (Organic)	Gawler, Australia	(Brookman, 2011)★	(AU$– US$ 0.95)	Farmers market	Peri-urban	$285,000
Four Season Farm	Maine, USA			Direct marketing	Urban	$294,643
SPIN intensive	Philadelphia, USA	(SPIN, n.d.)		Direct marketing	Urban	$360,000

Notes: ★ Brookman (2011), personal communication.
Based on single cropping. Multiple cropping would significantly increase return.

demand for expensive plant-based items. Although a higher risk than the more consistent demand for staple crops, it is potentially more lucrative, with concerns of oversupply or a lack of longevity able to be reduced following demand-based yield distribution research.

Les Jardins de la Grelinette is based on Coleman's (pers. com., 2014) approach and grosses $151,561/ha. In terms of labour intensity, two owners work full-time and are helped by one full-time farm hand from March to December and one part-time farm hand from June to November. Martin writes that owner 'Fortier asserts that a market garden less than 2.5 acres [1ha] can generate annual sales of $60,000 to $120,000, with a profit margin of more than 40 per cent' (2013: 35).

The 5,600m² Four Season Farm returns an impressive gross income from vegetable production equivalent to $294,643/ha. This environment has a limited growing season (Ableman, 2005) with challenging soil conditions requiring considerable labour-intensive and organic improvements (Coleman, pers. com., 2014). Coleman posits 'a farm that specialized in just the most-remunerative-per-square-foot crops could easily exceed our income figures' (ibid). After paying salaries and other expenses, net income is about $60,000 without mortgage or food costs (ibid). Four Season Farm's production system is based on four key approaches: (1) simplified production techniques; (2) using efficient small machinery and tools; (3) a system that reduces expenses on external inputs; and (4) marketing produce in a way that will bring the greatest return (Brown, 2008: 27). Coleman and Damrosch possess extensive expertise, experience and dedication and may be considered an exemplar of sustainable and productive small-scale agriculture.

The high intensity SPIN claims the greatest gross returns from any of the calculations at $360,000/ha, but requires high labour intensity. Based on pilot testing over four years, the Institute for Innovations in Local Farming projected economic feasibility to be possible after initial establishment – but at a slightly increased scale above the 0.5 acre (202m²) pilot area cultivated and through the clustering of small farms (Urban Partners, 2007: 2). Thus, the pilot did not achieve financial viability; in its fourth and most successful year, $68,000 gross was generated (equivalent to $336,070/ha) but operating costs exceeded income, being $69,600 ($51,200 for wages). The SPIN pilot emphasised the importance of direct marketing to maximise profitability, with their harvests sold via four market segments: four outdoor farmers markets (50 per cent of total sales); community supported agriculture (CSA) shares; restaurant/wholesale outlets; and an onsite farm stand (ibid). Other recommendations included clustering small intensive farms (four/six) to share infrastructure, knowledge and management of financial and legal interests. The public urban agriculture green space scenario proposed in this chapter could meet many of these recommendations and potentially exploit SPIN's efficiencies through economies of scale and suitable, economically efficient labour arrangements.

As demonstrated in Table 7.3, significant economic returns can be generated in public green spaces if urban agriculture is executed at 25 per cent or a more substantial 50 per cent of their existing area. Based on returns of $150,000 and $300,000/ha, if deployed at 25 per cent of the area for Central Park in New York this could gross $12,750,000–$25,500,000 annually. The same area percentage of

Table 7.3 Based on ha calculations (Table 7.2), gross return in US$ for 10– 2,500 hectares for each production method, with existing green spaces (Table 7.1) listed by >25 per cent and <25–>50 per cent of area and corresponding indicative $ return

Production area (ha)	10	25	50	75	100	150	200	250	350	500	750	1000	2500
USA All Vegetables (Organic)	$222,490	$556,225	$1,112,450	$1,668,675	$2,224,900	$3,337,350	$4,449,800	$5,562,250	$7,787,150	$11,124,500	$16,686,750	$22,249,000	$55,622,500
UK Yield & Retail Data: Onion (Conventional)	$351,640	$879,100	$1,758,200	$2,637,300	$3,516,400	$5,274,600	$7,032,800	$8,791,000	$12,307,400	$17,582,000	$26,373,000	$35,164,000	$87,910,000
AUST Farmers market: Apples (Organic)	$380,000	$950,000	$1,900,000	$2,850,000	$3,800,000	$5,700,000	$7,600,000	$9,500,000	$13,300,000	$19,000,000	$28,500,000	$38,000,000	$95,000,000
UK Yield & Retail Data: Apples (Conventional)	$412,930	$1,032,325	$2,064,650	$3,096,975	$4,129,300	$6,193,950	$8,258,600	$10,323,250	$14,452,550	$20,646,500	$30,969,750	$41,293,000	$103,232,500
USA Herbs (Organic)	$416,450	$1,041,125	$2,082,250	$3,123,375	$4,164,500	$6,246,750	$8,329,000	$10,411,250	$14,575,750	$20,822,500	$31,233,750	$41,645,000	$104,112,500
UK Yield & Retail Data: Carrot (Conventional)	$466,850	$1,167,125	$2,334,250	$3,501,375	$4,668,500	$7,002,750	$9,337,000	$11,671,250	$16,339,750	$23,342,500	$35,013,750	$46,685,000	$116,712,500
AUST Farmers market: Citrus (Organic)	$997,500	$2,493,750	$4,987,500	$7,481,250	$9,975,000	$14,962,500	$19,950,000	$24,937,500	$34,912,500	$49,875,000	$74,812,500	$99,750,000	$249,375,000
SPIN low value	$1,200,000	$3,000,000	$6,000,000	$9,000,000	$12,000,000	$24,000,000	$30,000,000	$42,000,000	$42,000,000	$60,000,000	$90,000,000	$120,000,000	$300,000,000
Fairview Gardens (Organic)	$1,458,333	$3,645,833	$7,291,665	$10,937,498	$14,483,330	$21,874,995	$29,166,660	$51,041,655	$51,041,655	$72,916,650	$109,374,975	$145,833,300	$364,583,250
Specialty crops	$1,482,660	$3,706,650	$7,413,300	$11,119,950	$14,826,600	$22,239,900	$29,653,200	$37,066,500	$51,893,100	$74,133,000	$111,199,500	$148,266,000	$370,665,000
Les Jardins de la Grelinette	$1,515,610	$3,789,025	$7,578,050	$11,367,075	$15,156,100	$22,734,150	$30,312,200	$37,890,250	$53,046,350	$75,780,500	$113,670,750	$151,561,000	$378,902,500
AUST Farmers market: Lettuces (Organic)	$2,850,000	$7,125,000	$14,250,000	$21,375,000	$28,500,000	$42,750,000	$71,250,000	$99,750,000	$99,750,000	$142,500,000	$213,750,000	$285,000,000	$712,500,000
Four Season Farm	$2,946,430	$7,366,075	$14,732,150	$22,098,225	$29,464,300	$44,196,450	$58,928,600	$73,660,750	$103,125,050	$147,321,500	$220,982,250	$294,643,000	$736,607,500
SPIN intensive	$3,600,000	$9,000,000	$18,000,000	$27,000,000	$36,000,000	$54,000,000	$72,000,000	$90,000,000	$126,000,000	$180,000,000	$270,000,000	$360,000,000	$900,000,000
>25% park area		Washington Park, Royal Park	Hyde Park	Central Park	Bushy Park, Englischer Garten	Phoenix Park, Adelaide Parklands	Bushy Park	Englischer Garten	Phoenix Park, Adelaide Parklands			Western Sydney Parklands	
<25–>50% park area				Washington Park, Royal Park	Hyde Park	Central Park						Western Sydney Parklands	

Per cent of park area (row group label)

Growing system (annual gross return US$)

the Adelaide Parklands could gross $28,500,000–$57,000,000, while if initiated for 50 per cent of the expansive Western Sydney Parklands, $400 million to $800 million could be reached. These figures are very basic projections and would require substantial deployments of skilled and experienced staff for growing, management and sales of produce and years of optimisation and experience. Nonetheless, Fairview Gardens, The Food Forest, Les Jardins de la Grelinette and Four Season Farm all provide existing successful models of economically viable private enterprises that could be adapted for public green spaces, with careful consideration of a range of factors ranging from maintenance, to growing seasons, aesthetics and public interface.

Additional considerations

Labour and maintenance

While existing green spaces already have maintenance staff employed for upkeep, in most cases (except for highly manicured spaces), urban agriculture requires increased labour (determined through site-specific research). As a guide, 2,000m^2 or 1/5th ha per individual is near the upper limit for intensive, low-mechanised farming as laid out by the Coleman, Martin and SPIN approaches (Coleman, 1995; Brown, 2008: 33). This equates to five staff/ha for intensive vegetables and herbs, which could gross around +/-$300,000/ha. Fruit and nuts are less labour intensive, as are some specialty crops. It is important to note that farming involves variable, long hours during the growing season and temporary additional labour may also be required at critical seasonal times for harvesting, processing and selling produce.

The public context and scale of urban agriculture hypothesised in this chapter can benefit from innovation (Mason & Knowd, 2010: 67-68) and research arrangements with industry and educational institutions. A number of approaches to labour structuring are possible to increase economic viability while contributing to social outreach and enrichment:

- Employment of supervisor/manager/officer to oversee and manage temporary staff;
- Student placements, apprenticeships, qualifications (these need to align with growing seasons);
- Volunteers;
- Payment for labour in produce;
- Labour sourced from unemployed work for benefits and/or prison labour and rehabilitation schemes;
- Work and industry placement schemes;
- Socially disadvantaged initiatives;
- WWOOF ('Willing Workers on Organic Farms' – usually provided with board and lodging);
- Public–private partnerships through in-house and contracted labour (as in the case of maintenance contracts by many municipal governments);

- Leasing of land to suitable organisations (experienced, not for profit, ethical, community sensitive models rooted in ethical and social approaches to sustainable practices) (Thompson, 2007);
- Corporate sponsored labour and sustainability schemes.

Management of these labour arrangements by experienced personnel is fundamental to success, with core-salaried and experienced staff crucial to continuous, smooth day-to-day operations.

Soils

In many cases, existing soils will not be optimised for urban agriculture or may be contaminated. This is not a significant barrier, and can be overcome by growing in compost-filled raised beds. Intensive growing techniques return the highest economic values, but need to maintain soil fertility. Very high amounts of compostable waste are available in urban contexts and these can provide both a growing medium and on-going fertility (see the economically viable, compostable waste growing systems in Chicago by Dunn (Finkel, 2012)).

Establishment

The costs of establishment need to be detailed on a site-specific basis. The initial costs of capital works will depend on the aesthetic standard desired, the extent of machinery, the number of buildings, the extent of green infrastructure and the scale of the intervention. Some of these costs may be subsidised through 'green', lottery and research funding mechanisms and industry and philanthropic sponsorship. Plant establishment additionally requires costing, as economic returns can be delayed by species that do not yield for several years (such as fruit and nut trees).

Growing season

Economic returns can be significantly affected by growing seasons, with some species yielding more than one crop per season. The growing season can be extended in colder locations through relatively inexpensive approaches such as poly-tunnels (as utilised at Four Season Farm) or walipinis (earth-sheltered, non-heated greenhouses). These can be aesthetically designed to fit public green spaces (Zeunert, 2011).

Produce processing and retailing

It is important to optimise sales of produce if urban agriculture is deployed at a medium to large scale. Gross economic returns were highly favourable for operations utilising direct marketing of produce through a range of techniques, including direct retail (for example, onsite), farmers markets, CSA arrangements (Sumner *et al*, 2010) and restaurant/wholesale outlets. Direct marketing in the hypothesised scenario is highly feasible due to the immediate proximity of significant volumes of buyers. Other value-adding opportunities include preserves, drinks, gourmet and ornamental products and crafts, and onsite café/restaurant outlets. Branding, marketing and identity can strengthen these endeavours (Mason & Knowd, 2010).

Additional income sources

Aquaculture, animal husbandry, apiaries and intensive greenhouses have not been calculated but could significantly increase income/ha (more challenging perhaps in public spaces, although the precedent has been set in various urban farms across developed countries). Animals, especially grazing animals and horses, have a long-standing relationship with public green spaces. The addition of chickens, ducks, rabbits or other suitable animals offers increased productivity and income (such as in the case of beehives generating both income from food production and productivity through pollination – essential to maximising the yield of many crops). Initial capital costs for infrastructure such as fencing must be considered, as well as zoning restrictions and bylaws, which may affect both built structures and/or the keeping of animals. Aquaculture can yield high returns relative to the space required but again incurs initial capital investment and on-going safeguarding and should be coupled with WSUD (water sensitive urban design) strategies to ensure optimum water cycle management. Intensive greenhouse production has been excluded due to the 'built' nature of greenhouses usually perceived to be inappropriate in public green space, although walipinis can greatly reduce visual impact and heating requirements. These could significantly raise production levels especially in colder climates (Cities, 2013: 149).

Social and educational activities offer further revenue streams, including workshops; education courses; business incubation; training; visits; tours; café/restaurants; qualifications; agro-tourism; and school visits and stays. Several of the existing growers examined (Fairview Gardens, The Food Forest) successfully utilise opportunities to supplement and in some cases, provide significant income. Additionally, philanthropy and donations provide further income, dependent on social and environmental initiatives, public relations, marketing and media coverage.

Additional costs and planning

Some jurisdictions provide exemptions and incentives for small growers, helping small-scale farming's economic feasibility. Costs may need to be factored for regulations and testing where beyond certain production limits, while planning, legislation and policy may require updating or adjustment in order to realise the visions discussed in this chapter.

Aesthetics and interface

Including urban agriculture within existing public green spaces requires careful consideration and does not need to be uninviting if the planning, design and execution are well considered (Zeunert, 2011). It is important that urban agriculture be perceived as an integral and fitting part of green space, especially at the public interface and edges. A range of design considerations includes: size/scale (horizontal/vertical); siting/views; interface/edges (enclosure/openness, hard/soft/blended); signage/language (hostile versus open); the extent of human activation/participation/presence in the landscape (vacant versus thriving); public/private (accessible as opposed to a business/commercial environment); and the aesthetic and visual performance, arrangement and planting composition.

Conflict

If urban agriculture were to be deployed in existing green spaces, examples of possible hindrances include: increased consultation and possibility for complaints and conflicts, increased policy and planning complexity, and potential for theft and vandalism. These are standard challenges in the planning and design of public spaces and can be appropriately mitigated and managed for optimum harmony and benefit through existing municipal mechanisms and procedures.

Conclusion

Urban agriculture is a legitimate green space activity in its own right. Only relatively recently marginalised, it has historically formed an integral part of civilised public life. Conversely, the more recent historical legacy of the ornamental picturesque continues to dominate, despite performing poorly when compared to productive, nuanced, multidimensional and participatory techniques that contemporary public green spaces can generate (Knight & Riggs, 2010; Yu, 2010). This emerging practice's economic viability, employment, local food production and educational, as well as expanded health and environmental benefits, need further research, exploration and testing if it is to realise its potential as a new paradigm for public green space.

When proficiently planned and managed, urban agriculture offers a civically aligned income source, or perhaps more importantly, a socially and environmentally productive use of public open green space. Private operations (such as Fairview Gardens, The Food Forest, Les Jardins de la Grelinette and Four Season Farm), demonstrate that gross returns of $150,000 to $300,000 (per annum/ha) are feasible and if these operations were applied in 25 per cent of the area for Central Park in New York this could gross $12,750,000 to $25,500,000 annually, in the Adelaide Parklands $28,500,000 to $57,000,000 and at 50 per cent of the Western Sydney Parklands, around $400 million to $800 million. To determine economic viability and potential net income, detailed and site-specific research is required to examine key considerations of intensity; scale; labour; initial and on-going costs; crop-selection; and maximising multiple opportunities for production. In the public domain and under proficient management, it is likely that these can be optimised to return significant economic and social benefits. 'Our desire for beauty detached from utility is weakening, and it should be. In our new world, survival is at stake. Wastefulness becomes viscerally unattractive, if not immoral. But there is plenty of opportunity for joyful pleasure in useful things' (Yu, 2010: 50).

Existing green spaces can be enshrined in protection and heritage legislation, with public sentiment counterproductive to environmental and economic sustainability. Targeting and unlocking lesser-activated areas is perhaps the first step. Risk-averse municipal authorities might be more responsive when presented with the multitude of economic, employment, social, community and health benefits, which would additionally engender training, research, business incubation and activation of local economies and networks. Current municipal processes which facilitate public consultation, the design process, on-going management and maintenance and ultimately, careful planning, would ensure the necessary navigation of a new paradigm of economically and socially generative public green space. Top-down buy-in from mayors and key decision makers would greatly enhance the potential for implementation.

Crucially, if we are to realise socially, civically and environmentally productive public green space, economic interests must be housed within an ethical framework – not just in theory but also in practice. In this vision, public green space can be recast in the spirit of 'the commons' for creating resources to provide multiple economic possibilities and community livelihoods, executed within a generative aesthetic. Should we wait for a food supply crisis to make our urban green spaces more productive, participatory and sustainable and our population more skilful in food production?

Acknowledgements

The author would like to thank Alys Daroy for detailed review and editing, Prof. David Jones, Dr Beau Beza and Tim Waterman for their assistance in the peer-review process, Prof. Rob Roggema for his work in organising the publication, and Graham and Annemarie Brookman (Food Forest, Gawler, Australia), Eliot Coleman (Four Season Farm) and Martin Cook (Adelaide City Council) for their assistance in providing economic data.

References

Abelman, M. (2005) *Fields of Plenty: A farmer's journey in search of real food and the people who grow it*, Chronicle Books: San Francisco.

Alon-Mozes, T. (2009) Landscape architecture and agriculture: Common seeds and diverging sprigs in Israeli practice, *Landscape Journal* 28(2): 166–180.

Berry, W. (1985) A few words in favor of Edward Abbey, *Whole Earth Review* (45)3: 38–44.

Blaine, T. W., Grewal, P. S., Dawes, A. & Snider, D. (2010) Profiling community gardeners. *Journal of Extension* 48(6): 1–12.

Brown, K. H. & Jameton, A. L. (2000) Public health implications of urban agriculture, *Journal of Public Health Policy* 21(1): 20–39.

Brown, S. (2008) Urban Agriculture: Is There Now An Opportunity For A Viable Smallscale Sustainable Agriculture To Emerge In Brisbane, Australia?, Masters Dissertation, Charles Sturt University.

Cabannes, Y. & Dubbeling, M. (2001) Urban Agriculture and urban planning: what should be taken into consideration to plan the city of the 21st century. In H. Hoffman & K.

Mathey (eds) *Urban Agriculture and Horticulture, the Linkage with Urban Planning*, International Symposium, Berlin.

CABE Space (2004). *Green Space Strategies: A good practice guide*. CABE Space: London.

Cities (2013) *Farming the City*, Transcity Valiz: The Netherlands.

Coleman, E. (1995) *The New Organic Grower*, Chelsea Green: Vermont.

Corner, J. (1992) Representation and landscape: Drawing and making the landscape medium, *Word & Image* 8 (3): 243–275.

DECC (Department of Energy and Climate Change) (2013) *Thermal Growing Season in Central England*, viewed 25 August 2014 <www.gov.uk/government/uploads/system/uploads/attachment_data/file/192601/thermal_growing_season_summary_report.pdf>

Doron, G. (2005) Urban agriculture: Small, medium, large, *Architectural Design* 75(3): 52–59.

Drayson, K. (2014) *Green Society: Policies to improve the UK's urban green spaces*, Policy Exchange: London.

Duchemin, E., Wegmuller, F., & Legault, A. M. (2008) *Urban Agriculture: Multidimensional tools for social development in poor neighborhoods*. Field actions science report, 1(1) viewed 15 August 2014 <http://factsreports.revues.org/index113.html>

Dunnett, N., Swanwick, C. & Woolley, H. (2002) *Improving Urban Parks, Play Areas and Green Spaces*, DTLR: London.

Enfield (no date) *Garden Enfield – About*, viewed 15 August 2014 <www.enfield.gov.uk/gardenenfieldMS/info/1/>

Finkel, E. (2012) *A Growthful Enterprise on Perry Street*, New Communities Program, viewed 18 May 2014 <www.newcommunities.org/news/articleDetail.asp?objectID=2434>

Firth, C., Maye, D. & Pearson, D. (2011) Developing 'community' in community gardens, *Local Environment* 16: 555–568.

Flores, H. C. (2006) *Food Not Lawns: How to turn your yard into a garden and your neighborhood into a community*, Chelsea Green: Virginia.

Foucault, M. & Rabinow, P. (1984) *The Foucault reader*, New York: Pantheon Books.

Foderaro, L. W. (2014) *Focusing on Lesser-Known Open Spaces in New York*, May 6, viewed 10 June 2014 <www.nytimes.com/2014/05/07/nyregion/focusing-on-lesser-known-open-spaces-in-new-york.html?_r=0>

Garnett, T. (1996) Farming the city: the potential of urban agriculture, *The Ecologist* 26(6): 299–307.

Golden, S. (2013) *Urban Agriculture Impacts: Social, Health, and Economic: A Literature Review*, UC Sustainable Agriculture Research and Education Program, viewed 18 June 2014 <http://asi.ucdavis.edu/resources/publications/UA%20Lit%20Review-%20Golden%20Reduced%2011-15.pdf>

Grewal, S. & Grewal, P. (2012) Can cities become self reliant in food? *Cities* 29: 1–11.

Halweil, B. (2004) *Eat Here: Reclaiming Homegrown Pleasures in a Global Supermarket*, W W Norton & Company: Washington DC.

Hassink, J., Zwartbol, C., Agricola, H., Elings, M. & Thissen, J. (2007) Current status and potential of care farms in the Netherlands, *NJAS Wageningen Journal of Life Sciences* 55: 21–36.

Kaplan, R. (1973) Some psychological benefits of gardening, *Environment and Behavior* 5(2): 145–162.

Knight, L. & Riggs, W. (2010) Nourishing urbanism: a case for a new urban paradigm. In C.Pearson, S.Pilgrim, J.Pretty (eds) *Urban Agriculture: Diverse activities and benefits for city society*, Earthscan: London.

LI (Landscape Institute), London Branch (2013) *Cultivating the City*, Dalston Eastern Curve Garden, London, 14 September 2013.

Lazarus, C. (2005) *Urban Agriculture: Join the revolution*, viewed 18 May 2014 <www.newvillage.net/Journal/Issue2/2urbanagriculture.html>

Lee-Smith, D. & Prain, G. (2006). Urban agriculture and health. In Hawkes C. & Ruel M. T. (eds) *Understanding the Links Between Agriculture and Health*, International Food Policy Research Institute: Washington, DC.

Leeuwen, E. van, Nijkamp, P. & Noronha-Vaz, T. de, (2010) *The Multifunctional Use of Urban Greenspace*, viewed 18 May 2014, <www.eukn.org/E_library/Urban_Environment/ Urban_Environment/The_multi_functional_use_of_urban_green_space>

Linn, C. (1999) *Reclaiming the Sacred Commons*, viewed 14 May 2014, <www.newvillage.net/assets/docs/linn.pdf>

Loukaitou-Sideris, A. (2006) *Urban Parks*, Southern California Environmental Report Card 2006, viewed 10 August 2014 <www.environment.ucla.edu/media/files/Urban-Parks-2006.pdf>

Malakoff, D. (1995) *What good is community greening?* Philadelphia, PA: American Community Gardening Association.

Maller, C. (2002) *Healthy Parks, Healthy People: The health benefits of contact with nature in a park context*, Deakin University: Melbourne.

Martin, J. (2013) Growing better, not bigger, *Small Farm Canada*, May–June 2013: 34–37, viewed 29 August 2014 <www.themarketgardener.com/wp-content/uploads/2013/10/ growingbetter_SFCmay20131.pdf>

Masi, B. (2008) Defining the urban-agrarian space, *Cities growing smaller*, Kent State University's Cleveland Urban Design Collaborative: 85–102.

Mason, D. (2006) *Urban Agriculture – to identify how sustainable urban agriculture can benefit the quality of life of Australian communities*, Report to the Winston Churchill Memorial Trust of Australia.

Mason, D. & Knowd, I. (2010) The emergence of urban agriculture, *International Journal of Agricultural Sustainability* 8(1): 62–71.

McClintock, N., Cooper, J., & Khandeshi, S. (2013) Assessing the potential contribution of vacant land to urban vegetable production and consumption in Oakland, California, *Landscape and Urban Planning* 111: 46–58.

Merson, J., Attwater, R., Ampt, P., Wildman, H., & Chapple, R. (2010) The challenges to urban agriculture in the Sydney basin and lower Blue Mountains region of Australia, *International Journal of Agricultural Sustainability* 8: 72–85.

Morgan, K. (2013) Feeding the city: the challenge of urban food planning, in *Cities Farming the City*, Transcity Valiz: The Netherlands.

Moustier, P. & Danso, G. (2006) Local economic development and marketing of urban produced food. In R. van Veenhuizen (ed.) *Cities Farming for the Future: Urban agriculture for green and productive cities*, RUAF: Ottawa.

NAO (National Audit Office) (2006) *Enhancing Urban Green Space, Report by the Comptroller and Auditor General*, viewed 30 July 2014 <www.nao.org.uk/wp-content/uploads/ 2006/03/0506935.pdf>

Nettle, C. (2010) Community gardening as social action: the Australian community gardening movement and repertoires for change, PhD Dissertation, viewed 28 July 2014 <http://digital.library.adelaide.edu.au/dspace/handle/2440/71174>

Office for National Statistics (2013) *Consumer Price Indices – Series*, viewed 28 July 2014, <www.ons.gov.uk/ons/datasets-and-tables/data-selector.html?table-id=3.1&dataset=mm23>

Patel, I. C. (1991) Gardening's socioeconomic impacts. *Journal of Extension* 29(4): 7–8.

Pollock, M. (2002) *The Royal Horticultural Society: Fruit and Vegetable Gardening*, DK: London.

Pudup, M. (2008) It takes a garden: Cultivating citizen-subjects in organized garden projects, *Geoforum* 39: 1228–1240.

Pretty, J. (2002) *Agri-Culture: Reconnecting People, Land and Nature*, Earthscan: London.

Profitable Plants Digest, viewed 17 October 2014

Sherer, P. (2006) *The Benefits of Parks: Why America Needs More City Parks and Open Space*, The Trust for Public Land, viewed 25 August 2014 <www.eastshorepark.org/benefits_of_parks per cent20tpl.pdf>

Sumner, J., Mair, H. & Nelson, E. (2010). Putting the culture back into agriculture: civic engagement, community and the celebration of local food, *International Journal of Agricultural Sustainability* 8: 54–61.

Sommers, P. & Smit, J. (1994). *Promoting Urban Agriculture: A strategy framework for planners in North America, Europe and Asia*, IDRC: Ottawa.

SPIN (no date) Gardening Calculator Passalong, viewed 10 August 2014 <www.spingardening.com/common/pdfs/SPIN_Gardening_passalong_calculator.pdf>

SPUR (2012) *Public Harvest*, SPUR Report: 1–36.

St. Martin, K. (2009) Toward a cartography of the commons: Constituting the political and economic possibilities of place, *The Professional Geographer* 61(4): 493–507.

Thompson, I. (2007) The ethics of sustainability. In: J. Benson, & M. Roe (eds). *Landscape and Sustainability*, Routledge: London.

Urban Partners (2007) Farming in Philadelphia: Feasibility Analysis and Next Steps, Prepared for Institute for Innovations in Local Farming, viewed 19 August 2014 <www.spinfarming.com/common/pdfs/STF_inst_for_innovations_dec07.pdf>

USDA (United States Department of Agriculture) (2008) *2007 Census of Agriculture: 2008 Organic Production Survey*, National Agricultural Statistics Service, October 2012, viewed 9 August 2014 <www.agcensus.usda.gov/Publications/2007/Online_Highlights/Fact_Sheets/Practices/organics.pdf>

USDA (2012) *2011 Certified Organic Production Survey*, National Agricultural Statistics Service, October 2012, viewed 9 August 2014 <http://usda01.library.cornell.edu/usda/current/OrganicProduction/OrganicProduction-10-04-2012.pdf>

Wallin, C. (no date) 10 Most Profitable Specialty Crops to Grow, viewed 10 October 2014 <www.profitableplantsdigest.com/10-most-profitable-specialty-crops-to-grow/>

Woolley, H. (2004) *The Value of Public Space: How high quality parks and public spaces create economic, social and environmental value*, CABE: London.

Yu, K. (2010) Beautiful big feet: Toward a new landscape aesthetic, *Harvard Design Magazine* Fall/Winter 2009/10: 48–59.

Zeunert, J. (2011) *Eating the Landscape: Aesthetic Foodscape Design and its role in Australian Landscape Architecture*, AILA National Conference, viewed 22 August 2014 <https://joshuazeunert.files.wordpress.com/2011/08/aila_transform_afd_jz_revb.pdf>

Zeunert, J. (2013) Challenging assumptions in urban restoration ecology, *Landscape Journal* 32(2): 231–242.

8 Urban agri-tecture

The natural way of smart living

Natalia Mylonaki, with Bauke de Vries, Maarten Willems and Tom Veeger

Introduction

The pace of life in modern cities has significantly changed during recent decades. The increasing amount of working hours, new technological advances and many other factors have played an important role in this change. People are gradually coming closer and adopting new technologies that promise to leverage their quality of life, while the physical communication between them is tending to decline. Family members tend to be attracted to their personal screens (Liz & Revoir, 2010; Stelter, 2009) and the virtual environment of social media, a behaviour that enhances family alienation and is also supported by other aspects of the new way of life in cities.

Another important aspect regarding the quality of life in the cities is the changes that have been recorded during recent years in the nutrition habits of their residents. A meal has always been considered not only as vital calorie intake, but also as a social event. Even though it is impossible to set a general rule for all the developed countries of the world, a common tendency is readily recognizable. In the past, families used to gather around the table for lunch and dinner, while at the same time they had the opportunity to express their feelings and share their problems. Meals were made by the mother as a token of love and the ingredients were mostly grown by the family. This assured a respect and a deep knowledge of the identity of each meal. Nowadays, however, family members tend to have fewer meals together, especially on weekdays, the meals are often pre-cooked and heated to be served, and most of the time the origins of ingredients are unknown to them. However, annual growth rates of 16 per cent in Europe and 11 per cent in the United States in the sale of organic food and drinks may indicate a change towards a healthier behaviour with respect to our nutrition (De Borja et al., 2010).

With these facts regarding the family and communal alienation and the significant change in our nutrition habits in mind, a question is inevitably raised: is it actually possible to improve the interpersonal relationships in the urban family by recalling the old nutrition habits and, if it is, how can innovative technologies be used in order to achieve this goal?

Urban agriculture as a way of smart living

Nowadays, the food needs for the urban population are higher than in the past and the matter of nutrition in the cities will be increasingly discussed in the future. Undoubtedly, diet and nutrition have clear health linkages. A diet poor in vegetables and fruit is associated with cardiovascular diseases, certain types of cancer, micronutrient deficiencies, hypertension, anemia, premature delivery, low birth weight, obesity and diabetes (Pederson & Robertson, 2001). However, the high demand for fruit and vegetables from cities around the world requires massive greenhouse cultivation and long-distance transportation, which often leads to the deterioration of their quality and nutrients.

A potential solution to this problem has recently been suggested in the form of urban agriculture, which involves bringing food production closer to or in the cities themselves. This method is likely to provide urban residents with fresh fruit and vegetables of higher nutritional value than those that are stored and transported for a long time. It is strongly argued that 30 per cent of the food could be produced within the city borders, with significant benefits to its residents and the city environment itself (Whitfield, 2009).

This chapter focuses mostly on the social benefits of urban agriculture, which is considered to be a highly communally driven and managed activity. Its strategies are commonly used for the social integration of newcomers and underprivileged groups in the urban socio-economic system. Moreover, people of different ethnicities, genders, classes and ages come together towards a common goal. In this way, stronger urban communities with significant collective identity are built within the city limits. Furthermore, abandoned plots and dump sites that were used for illegal activities are now being used for growing food activities that make people come closer to each other, interact and communicate. Another aspect that has to be mentioned is the educational value of urban agriculture. As people come closer to production, they gain knowledge of the origins, cultivation techniques and the proper growing season of what they eat. Last but not least, involvement in growing food has psychological benefits too. Many people in developed countries admit that they do not undertake this activity for food production itself, but for the fact that it helps them to relieve their stress, so they consider it as a healthy pastime.

Taking the high importance of green policies in construction and the ever-growing popularity of urban agriculture into consideration, developers are likely to invest in such activities. The most commonly proposed architectural solutions are community gardens, green roofs, living walls and the latest idea of vertical farms. As far as technology is concerned, many new systems have been developed in recent years in order to support urban agricultural activities. Most of them are based on new technologies that aim to make the growing of food possible in ways other than the ordinary. Concerning growing plants, hydroponics, aquaponics and aeroponics are emerging production technologies that do not use soil as a medium. Another interesting method named fungiculture is used for growing mushrooms indoors with the use of straw filled bags. Apart from growing plants, some methods for growing animals have also been developed. Urban beekeeping, fish farming and heliciculture (snail farming) are activities that are gaining more and more fans every year in big cities like New York and London.

It is safe to claim that technological advances have contributed to family alienation, however, technology could be used in order to tackle this problem through urban agriculture activities. Technology in this domain is highly developed and cultivation has thus become easier, more accessible to city residents and compact enough so as to be fitted even at home. As such, putting it into practice may only bring about positive results towards the improvement of city residents' quality of life.

The thesis of this chapter is that new activities around food can contribute to the reinstatement of old healthy habits and bring people together towards a common goal. Through the process of growing food, people will interact and the knowledge behind their meals will be regained. Moreover, the yield of fresh and healthy food will be rewarding and good for their health.

Project description

To date, the urban agriculture movement has focused primarily on how cities could become self-sufficient with respect to food supplies and on solving the lack of food supplies in developing countries. In a critical review of this goal, I strongly believe that this is a utopian view of the subject as the means to achieve such self-sufficiency are both expensive and difficult to obtain. Essentially, this is the reason why these futuristic projects have not yet been implemented. Nonetheless, if we focus on the main idea that this movement proposes, we can take advantage of its high social and nutritional benefits. The systems used to support the urban agriculture are mostly developed for achieving mass production. However, as these changes are more likely to occur firstly in the home context, the issue is how these innovative technologies could be implemented in a creative and sustainable way.

The project is structured with a view to achieving the two main intended results. The first is the implementation of small and bigger scale activities of urban agriculture in the urban house with the use of new technological advances. The second is the provision of a better quality of life for residents of cities in two dimensions: psychologically, by the improvement of the social life of the residents, and physically, by the provision of healthier nutrition. In that way, urban houses will provide their residents with fresh food and encourage them to develop better interpersonal relationships at family and communal levels.

As far as the building typology is concerned, the project proposes a new design for the Dutch urban row house. As the row housing typology is widely preferred in the Netherlands (Heijneman & Ham, 2004), the proposal of an alternative design based on the urban agriculture philosophy may achieve the sustainability of future buildings by its broad potential application in the future housing system. Moreover, it is common that allotment gardens that are rented to urban residents are developed outside the cities. This project dares to bring this habit in to the urban environment and in the house itself.

In order to thoroughly comprehend the values of the new row house design, the case study of Woensel-West in Eindhoven was chosen due to the social conditions, the multicultural context of the area and the existing strong community. The area's basic characteristic is that it is inhabited by 79 different nationalities. As far as the building environment is concerned, most of the houses are family houses, built before the Second World War. Besides the buildings' physical problems, the social welfare of a group of the inhabitants is weak and safety has been an issue for many years. Over time, many plans have been implemented for the development of the area. The new way of living that is introduced by this project can enhance the already existing tendency of communal living and sharing, set the basis for a new orientation towards a safer neighbourhood, and contribute to the development of a green core that is missing from the area.

The concept

The site of the project is located in the heart of Woensel-West and it borders the Celsiusstraat, Brugmanstraat, Fahrenheitstraat Streets and the Celsiusplein Square. Its position in the area is important as it lies at the end of the Edisonstraat Street, one of the most important streets in the area. Although Celsiusplein is one of few squares in the area, it is used as a playground and its position is crucial, but it is nevertheless rundown. For this reason, the project includes the redesign of the square too.

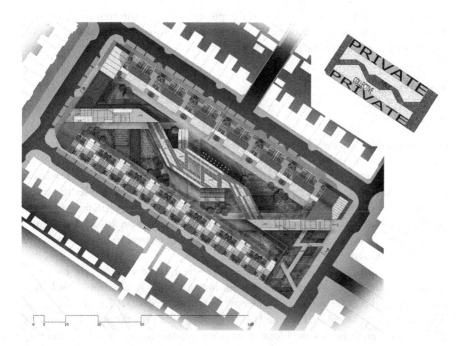

Figure 8.1 The integration of the project in the area and the top view of the block

The project has three scales of design: the block, the housing unit and the technological systems used. The block scale describes the complete design of a block and its main objective is to create a communal and a public core that mutually contribute to its development. The housing unit is designed upon the principles of the row-housing typology and can be built in every city in the Netherlands and abroad. Finally, the technological systems scale aims at the creative implementation of existing urban agriculture systems in the design. Urban agriculture activities appear in all the designed scales. The users take advantage of growing and eating their own fruit and vegetables, coming together and developing a community under the scope of agriculture.

In order to make the idea of living in an agricultural environment even more appealing, the activities around agriculture do not burden the users financially, as all scales are designed in such a way that they can provide their users with self-sufficient and sustainable production. For that reason, the idea of energy circles is introduced.

All scales are equipped with a system of areas that is called 'the battery', as the respective design is in fact based upon the working principles of a real battery; it produces and saves all the energy needed for urban agriculture systems to work and it delivers in the ways and times required. The mechanical systems that support the agriculture activities at all scales are placed in these areas, from now on referred to as 'batteries', which are interconnected in order to achieve optimal energy provision for the project. The design of the batteries is described below with reference to the three scales of the project.

In order to achieve self-sufficiency, the project employs the following systems: aquaponics, hydroponics, aeroponics, rainwater harvesting, wastewater treatment, waste composting and solar panels.

The design

The block scale

Conventionally, the block is developed in the typical enclosed Dutch way and has 34 houses with gardens. The new design proposes a block of 25 housing units where the public (>1000 people), the collective (>100 people, the block residents) and the private sectors (2–5 people, the houses' residents) function under the scope of urban agriculture. These sectors come together but are also separate in order to serve user needs best. The design of an open block was chosen as social awareness is accomplished by means of a route through it. Living or passing by the block will be a new experience in the urban fabric.

The basic concept of the block scale is the creation of a pedestrian road that crosses the block as an extension of the Edisonstraat, with all functions developing on and around it. The public and collective batteries are designed across this street, while the houses are placed on the two long sides facing northeast and southwest. The main characteristic of the block strategy is the interaction of residents and visitors. The level difference is crucial to achieve two simultaneous circulations in two

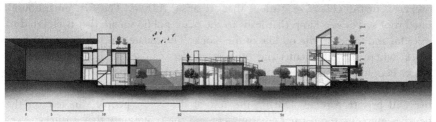

Figure 8.2 From top to bottom: the collective sector; the A–A section of the block where the all scales are seen; the public sector; the B–B section of the block where the connection of the public route and the collective greenhouse is shown

directions that can support not only visual but also physical connections when desired. The house units are designed in a strict way so as to be able to use the same design in other blocks too, while the public sector is freer and site specific.

The public sector

The public road forms a route through the block that brings visitors closer to the movement of urban agriculture. Its shape is based on the orientation of the greenhouses and follows a crooked pattern, with a view to offering different views for the visitors that walk on top of them. The starting point of the route is the educational centre and the garden that is placed at the beginning of Fahrenheitstraat, on level +0.00 metres. The route continues with a stepped ramp that leads to the first elevated observation point in +2.10m. At that point, the visitors can have a first overview of the collective gardens that are at at -1.00m. The first greenhouse's tilted roof serves as a ramp that leads to the main public square that is elevated to +3.50m and stands in the middle of the block. The square hosts the market of the block, where the inhabitants and other producers can sell their products once a week. People can buy the fresh products while in the location of production, namely on the two big collective greenhouses and surrounded by open-air collective and private gardens. During the rest of the week, people can rest on the benches enjoying the natural environment and the tranquillity that it offers.

The route continues over the fourth collective greenhouse, towards the second observation point in +2.10m. The route ends at +0.00m, between the entrance to the public restaurant and the multifunctional open space that can be used as a relaxing area, a playground or an outdoor space used in public events with the support of the restaurant.

The collective sector

Set at -1.00m, the collective area consists of greenhouses, storage rooms, gathering points, gardens and water tanks. The buildings under the public stepped ramps and the glass and metal structure that runs through the block and hosts a major part of the collective sector make up the so-called collective battery, with the mechanical systems that support production and the plant rooms placed in it. The plant room of the systems used in the area and a storage place lie under the eastern stepped ramp. The two small collective greenhouses with the tilted roofs host low vegetation, while the two central ones can host mid-sized trees that cannot survive outside and the collective aquaponics system. The western stepped ramp hosts collective fungiculture and the production storage room and has a direct connection to the market. All these covered areas have small gathering hubs where the people living in the block can meet, eat and enjoy their leisure time.

With respect to the open-air part of the collective sector, gardens, paths and water tanks are designed. Every garden is cultivated with a specific type of vegetation depending on the period of year and it is taken care of by the inhabitants of the block. Moreover, a plot is dedicated to the activity of urban beekeeping. Last but not least, the water tanks installed in between the gardens are part of a rainwater harvesting system that is set in the plant room. They have a linear design of 1m depth, so as to avoid high levels of evaporation. They are placed in partly shadowed areas where crops and vegetables would be difficult to grow and next to the greenhouse, so that they can get the water from the roofs too.

The block's circles and systems

In order to achieve self-sufficiency in block production, the public and the communal batteries are supplementary and interconnected. Solar panels are installed on the roofs of the educational centre and the restaurant, while the rainwater that is collected in the water tanks and the wastewater from both sectors is transferred in the plant room and is used for irrigation in the collective and public sectors. Finally, the organic waste produced in all activities is composted and used as fertilizer in the open-air gardens and in the systems that depend on soil.

Concerning the way that food grows in the block scale, the greenhouses host aquaponic systems, while regular growing techniques are also used. Ventilation is provided by natural means through openings on the top of the glass walls, but also mechanically by means of the installed ventilation system. Moreover, under the steel plates of the roof hangs the lighting system that is needed for the days with limited sunlight. For the open-air gardens, temporary semi-transparent structures are designed in order to protect the plants from unfavourable weather conditions.

The housing unit scale

The project proposes an alternative design for the row-house typology that can be potentially implemented in every city. The rationale pervading the design lies on the satisfaction of the inhabitants' need to come together, discuss while eating and get fulfillment by growing their own food.

As the project revolves around urban agriculture activities, house orientation and light permeation play a very important role in their design. For that reason, it was necessary to develop two different designs that would share the same principles and still correspond to orientation demands in the best possible way. Type A is for houses with entrances to the north, while type B is designed for the south-oriented houses. Moreover, the two house units have different target groups. The ones in Celciusstraat (type B) are designed for couples or families with a maximum of one child, while those in Fahrenheitstraat (type A) are for families with two children or more. In this way, diversity is achieved both in the image of the block and in the types of people that use the collective areas. The units are mirrored and placed in pairs, one next to the other in the block borders.

The unit areas

The two types are developed by the synthesis of eating, leisure, relaxation/working, sleeping areas and the battery. The eating area is the core of the house design,

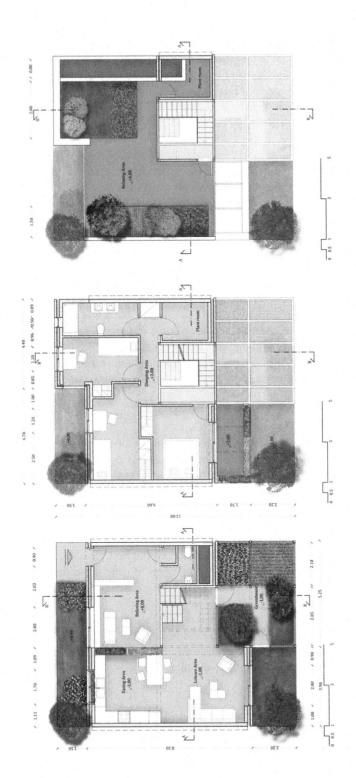

Figure 8.3 The ground, first and second floor plans

as it provides the main access to fresh food and contributes to family bonding. In order to enhance this value, the eating area is equipped with a special structure, which serves the cut-and-eat principle in the house and hosts the part of the aquaponic system where plants grow. The leisure area houses entertainment and educational functions. It is the place where the fish tank of the aquaponics system is installed as part of the battery and a smart wall that is connected with the agricultural system data and the cable television stands. In this way, the family members can always be updated with the status of their farming systems, be informed about the seasonal vegetables and, at the same time, enjoy their free time in a modern way. These two areas form the social core of the unit and, for this reason, they are placed next to the greenhouse and are immersed at -1.00m, the level of the collective garden. The relaxation area is separated from the social core of the unit, with a view to achieving the required peace and quiet and is split over the ground floor (+0.00m) and the terrace where a roof garden is designed (+6.00m). The sleeping areas are also isolated and shadowed and are placed on the first floor (+3.00m). In their balconies, an exterior version of the cut and eat structure is installed to ensure the safeguarding of the residents' privacy and the shading of the premises.

The battery hosts the plant rooms, the system pipes, the sanitary areas, the atrium and the unit's small private greenhouse. The plant rooms host the mechanical parts of all the systems that are installed in the house. As for the sanitary areas, a small toilet is placed next to the entrance, while the family bathroom is located on the first floor. The greenhouse and the atrium are two linked elements of the battery. The atrium allows for the circulation within the house premises, as the stairs are part of it. It is covered with U-channel glass that filters solar radiation and protects the users' privacy. The greenhouse is placed at the level of the social core (-1.00m) and has three functions. First, it provides the house with a protected area

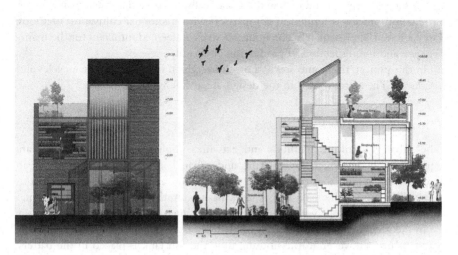

Figure 8.4 The south elevation and the B–B section of type A houses

for the growing of the private cultivations. With a height of 4m and covered with transparent glass, it can host medium-sized trees and other vegetables that are impossible to grow in the small-scale interior systems. Second, as the temperature in a greenhouse with a south façade can build up to 10 C while the outside temperature is 0 C, openings are designed in the glass walls in order to transfer the warm air to the interior. For ventilation needs, the greenhouse also has openings to the open air and a mechanical system is also installed. Finally, as it is placed next to the gardens, it can also be used as a covered relaxation area for the inhabitants. As the design of the houses is identical, part of the battery is exposed to the street, so that the vegetation growing in the batteries of each house can differentiate their façades.

The housing unit circles and systems

At the housing unit scale, everything is connected to the battery, either visibly or not and performs in energy circles. The smart wall in the leisure area and the systems installed in the battery core control the aquaponics systems. Moreover, all the organic waste is disposed of in the composting system attached to the battery next to the eating area. The fertilizer produced is used in the greenhouse and in the open-air gardens. The rainwater that is collected in the roof tank, as well as the wastewater produced in the sanitary and eating areas, is cleaned in the battery and used for irrigation. Finally, the solar panels installed on the roof produce sufficient energy for the systems to operate.

The systems scale

Bearing in mind the hectic everyday life rhythms in the urban context, technology is used as a tool to help the users in their urban agriculture activities. As all the systems are automated and operated mechanically, the time that users spend on the production process is limited in comparison with traditional cultivation methods. For each user, approximately one hour per week is deemed sufficient for the maintenance and the inspection of production at all scales.

In this project, the home acts as a machine, which filters, processes, recycles and produces. The systems used in the design are as follows.

Urban agriculture methods and activities

The urban agriculture methods and activities implemented in the design are aquaponics, hydroponics, fungiculture and urban beekeeping. Aquaponics is an autonomous system that consists of a fish tank and growing beds on top. In this project it is used as the main way of growing food in the eating areas, the balconies and as a supplementary way of growing food in the eastern collective greenhouse.

In the greenhouses, the system is implemented according to the relevant industry standards. However, in the housing unit, the fish tank is placed in the battery, while the growing beds are placed in the cut and eat structure. This is an original

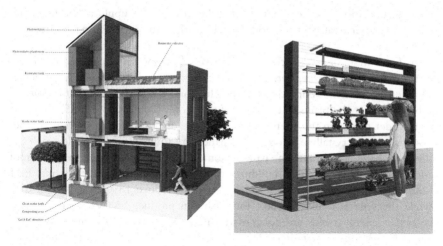

Figure 8.5 The system's location in the battery of type A houses and the cut and eat structure

design made for this project that consists of a metal structure, growing beds, irrigation pipes and the mechanical system. The growing beds are made of plastic, filled with gravel and soil, and they contain the typical irrigation system of aquaponics and are covered with the same wood used for the covering of the battery core, so as to achieve coherence in the final design. The first wooden box on top contains the water tank and the irrigation mechanism. The pipe that runs on the ceiling and comes from the fish tank fills the water tank. A pipe distributes the water to secondary pipes that are placed under the wooden boxes, so that fresh water is sprayed on the vegetables. The water that is taken out of the growing beds is transferred immediately back to the fish tank in order for the aquaponics circle to close. As far as lighting is concerned, having the Plant Lab prototype as a reference, the structures in the eating area are equipped with special LED lights that are installed under each box with a view to enhancing natural light and accelerating the growing process (Plantlab, n.d).

As for the other methods, traditional beekeeping techniques are used in a plot at the north side of the eastern greenhouse and fungiculture takes place in the area under the western stepped ramp. A dark and humid room is necessary in order for the mushrooms to thrive by the use of the straw bag technique.

Sustainable energy methods

As the project is established in the Netherlands, it would have been a true omission not to implement a rainwater harvesting system. The two housing types are equipped with water tanks of $4m^3$ each, while the collective water tanks have a capacity of $400m^3$ of water in total. The water is cleaned with the use of filters and used for irrigation purposes, but also for the fresh water supply in the fish tanks. The wastewater treatment method that cleans the grey water produced from the sanitary and eating areas is also used for the same purposes.

As far as waste disposal is concerned, in-vessel composting units are installed. In this way, needed fertilizer is produced and the composting of organic waste generates methane, which is burned to generate heat and CO_2. The CO_2 that is captured is beneficial for the plants in the aquaponic systems (Lennartsson, 2005).

In order to obtain the energy needed for the operation of the aforementioned systems, mono-crystalline photovoltaic panels are installed in the angle of 30°. In the case of the house units, the panels are mounted on a frame on top of the roof, and in the case of the public buildings, they are integrated into the roofing. Each house of type A can take nine panels that can produce 1890kWh/year in total, while the type B house can take four that can produce 85kWh/year. As far as collective energy needs are concerned, there are 28 panels on the educational centre and 34 on the restaurant that can produce 13.020kWh/year in total. These levels are regarded as sufficient for the mechanical systems' needs.

The systems that are used in the project are carefully chosen in order to meet its needs and are installed in a way that can only help and not hinder the users' everyday lifes. All their mechanical parts are installed in the batteries. Based on the energy circles, the systems of each unit are developed and interconnected with the others, ensuring that the energy produced is not wasted. The excess energy is stored in the batteries and is used when needed.

Conclusion

Urban agriculture is widely recognized as a way of achieving urban regeneration and its benefits to city inhabitants' nutrition and quality of life are highly instrumental. As stated in the book *Continuous Productive Urban Landscapes*, 'food growing projects can act as a focus for the community to come together, generate a sense of "can-do", and also help create a sense of local distinctiveness – a sense that each particular place, however ordinary, is unique and has value' (Viljoen *et al.*, 2005, p. 57). In the same book, urban agriculture is suggested as a way of tackling crime and reducing discrimination.

The case study of the underprivileged and multicultural area of Woensel-West was chosen for that specific reason – in order to support and promote the rehabilitation plans that are currently being developed by the Municipality of Eindhoven. A design of a new row-house typology was put forward, but the project also proposed a complete plan of the rehabilitation of the most central block in the area as a starting point for a whole new way of living.

The inhabitants and visitors of this block are given the chance to meet, talk and socialize, while taking part in all urban agriculture activities, from education, to the act of farming and eating, in both family and community contexts. Ideally, they would be prepared to adopt the proposed lifestyle, however, a sensible use of urban agriculture is suggested in order for the project to be feasible and attractive. Innovative technological systems are designed and employed in a creative way with a view to playing an assistant role in the production of food, contributing to a better, user-friendly design. For these reasons, this project aspired to answer the research questions that were stated in the beginning of this chapter.

It is worth pointing out that this was only a first attempt to create a neighbourhood based on urban agriculture in so many ways. This project can serve as the starting point for further discussion and proposals could be developed that work at even in a bigger scale than the single block. Regarding the social part of the project, an intense social study of a particular neighbourhood could be made. As far as technology is concerned, another design that could implement more types of urban agriculture may be considered. Of course, the systems and connections used in this project could also be altered or developed. Finally, as finances were not taken into consideration in this chapter, a SWOT (strengths, weaknesses, opportunities, threats) analysis and a complete business plan should be conducted for realistic implementation.

References

De Borja, H., Kuijer, L. & Aprile W. A. (2010). Designing for sustainable food practices in the home. Paper presented at Knowledge Collaboration &Learning for Sustainable Innovation ERSCP-EMSU conference. Delft, The Netherlands.

Heijneman, R. & Ham, M. (2004). Upgrading of post-war Row Houses in The Netherlands on the area of usable area and thermal resistance. Paper presented at Plea2004 – The 21th Conference on Passive and Low Energy Architecture. Eindhoven, The Netherlands.

Lennartsson, M. (2005). Recycling systems at the urban scale. In Viljoen, A., Bohn, K. &Howe, J. (eds) (2005). *Continuous productive urban landscapes: designing urban agriculture for sustainable cities* (pp. 89–92). Oxford: Architectural Press.

Liz, T. & Revoir, P. (2010). Computers and TV take up half our lives as we spend seven hours a day using technology. *Daily Mail*, 19 August. Retrieved 30 November 2012 from www.dailymail.co.uk/news/article-1304266/We-spend-7-hours-day-using-technology-computers-TV-lives.html#ixzz2EaYlnhq4.

Nelson+Pade Aquaponic technology, system and sypplies. (no date). *Aquaponics Information*. Retrieved 15 November 2012 from http://aquaponics.com/page/aquaponics-information.

Pederson, R. M. & A. Robertson. (2001). Food policies are essential for healthy cities. *Urban Agriculture Magazine (UA Magazine)*, 1(3): 9–11. www.ruaf.org/food-policies-are-essential-healthy-cities.

Plantlab. (no date). Revolution in plant growing. Retrieved 20 May 2013 from www.plantlab.nl/4.0/index.php/revolution-in-growing/.

Stelter, B. (2009). 8 hours a day spent on screens, study finds. *New York Times*, 26 March. Retrieved 30 November 2012 from www.nytimes.com/2009/03/27/business/media/27adco.html.

Van Veenhuizen, R. (ed.) (2006). *Cities Farming for the Future. Urban Agriculture for Green and Productive Cities*, Philippines: International Institute of Rural Reconstruction and ETC Urban Agriculture.

Viljoen, A., Bohn, K. & Howe, J. (eds) (2005). *Continuous Productive Urban Landscapes: Designing urban agriculture for sustainable cities.* Oxford: Architectural Press.

Whitfield, J. (2009). Seeds for edible city architecture. *NATURE, 459*, 18 June, pp. 914–915.

9 Urban waters for urban agriculture

ROOF WATER-FARMs as participatory and multifunctional infrastructures

Angela Million, Grit Bürgow and Anja Steglich

Introduction

In today's urban age, boundaries between cities and the countryside, between rural and urban, have been already blurring over the past decades. Today, agriculture comes into town and city life moves out into the hinterland. Likewise, existing water infrastructures for supply and disposal in many European cities and regions are at stake: in urban and rural areas there are challenges of maintaining and retrofitting existing central systems and transforming parts of them into more efficient and multifunctional infrastructure systems (Libbe & Beckmann 2010).

In Germany this challenge is accompanied by a number of research initiatives to explore, in theory and built practice, the structural transformations of linear centralized infrastructures towards decentralized approaches, integrated urban water management and design (e.g. BMBF 2012, Hoyer *et al.* 2011, AECOM 2013, DIFU 2013). A common aim is to promote the efficient usage of water resources to meet the challenges of demographic developments and climate change in order to enhance water security, both in supply and disposal. In addition, infrastructure systems are combined with other uses in urban everyday life (e.g. recreation, employment, education and food production) more than they have been before.

Interdisciplinary and transdisciplinary research is needed to explore these linkages of uses in water infrastructure (BMBF 2012: 6). This is especially necessary when explicit challenges are taken into account, such as the linkage of water and food supplies (BMBF 2012: 11). As water and soil, for example, are limited yet needed for food production, the development of innovative concepts and technologies are searched for – not only but also for an urban context. And what seems like major innovation and an abundance of technology has been the practice of combined water and food production in the global South (e.g. traditions of integrated aquaculture farming in Asia and South America) (Bürgow 2014). At the same time, contemporary international developments of an Urban Aquaculture arise. While originally, the focus was on water-farming, todays urban aquaculture also embraces facets of a water-living and water-wellbeing culture. It thereby reflects on the emerging need and trend of integrating multifunctional blue-green infrastructure typologies from swimming gardens to roof water-farm greenhouses into the daily urban realm (Ibid.). The question is whether this approach can be successfully adapted to European metropolises, their urban development and design, and urban lifestyles. This is explored by the interdisciplinary research project ROOF WATER-FARM,[1] funded from 2013 to 2016 by the Federal Ministry for Education and Research (BMBF) within the programme Smart and Multifunctional Infrastructure Systems for Sustainable Water Supply, Sanitation, and Stormwater Management (INIS).[2]

ROOF WATER-FARM[3] researched how wastewater can be a resource for urban food production via hydroponics (water-based plant cultivation) and aquaponics (combined fish and plant production). It is an onsite approach combining urban water management with urban farming (Bürgow *et al.* 2015). The building-integrated combination of water treatment (grey water and black water) with the production of fish and plants is a spatial and resource-efficient urban design strategy that uses the spatial potential of buildings' roofs within the European city context. Furthermore, a central aspect of the research is the tools and methods to incorporate the participation of different stakeholders, as well as education to foster a sustainable implementation of the ROOF WATER-FARMs (Bürgow *et al.* 2015).

Research framework and benefits of water farming and multifunctional infrastructure

Sustainability has increasingly been seen as an overarching mission statement for a steadily growing number of German communities. In relation to this political aim, a national research agenda, Research for Sustainable Development (FONA), has been set. One focus of the research is sustainable water management in relation to energy, food, health and the environment (BMBF 2012). The INIS research programme explores the future development of urban water management systems via multifunctional infrastructure (INIS) (DIFU 2013).

The urban vision of this programme is the transformation of monofunctional wastewater infrastructure to multifunctional infrastructures, described by terms such as ReUse, resource efficiency, local value chains, green infrastructure, Green City (EU 2010) and creative industries (Overmeyer *et al.* 2014). Here, the relation and positive synergies between blue infrastructure (water, sewage) and green infrastructure (open spaces, vegetation) is explored in academic and practice-oriented research. The approaches of multifunctional (blue-green) infrastructure and building-integrated water farming (Bürgow 2014) relate to theoretical theories of sustainable urban development or green urbanism in general (e.g. Beatley 2000, Barton *et al.* 2010), and specifically towards arguments for regenerative cities (WFC & HCU 2010) or loop cities (Verbücheln *et al.* 2013).

Regenerative cities and loop cities describe theoretical and possible resource cycles within the city. They ask for a more sustainable urban development and design that increases healthy synergies and interacting feedback cycles within the urban context for essential resources, such as energy, waste, biodiversity, water, food, soil, space and knowledge. The aim is to use resources more efficiently and to minimize environmental overreliance (Verbücheln *et al.* 2013). They are based on a model that aims to organize and develop cities as ecosystems (e.g. Todd and Todd 1993, Lyle 1994, Barton *et al.* 2010) and to look at them 'through the lenses of urban metabolism, ecological footprints, or eco-cycle balancing' (Beatley 2000: 416).

The model of resource metabolism – which aims for urban resource flows – is not new and has been part of the environmental debate for 20 years already. Nevertheless, a number of issues have not been adequately studied, among others, how environmentally friendly and consumer-friendly infrastructure systems will shape urban space and urban life in the future. The research approach of the Roof Water-Farm project is embedded into this framework. It particularly faces an urban aquaculture respectively water-farm based design as a building-integrated farming strategy and multifunctional blue-green infrastructure concept, which is lightweight, space- and resource-efficient (Bürgow 2014). In addition to, participatory planning, creative and self-made city discourses are an inevitable part of this approach (Lefebvre 2016; Ferguson 2014; Overmeyer *et al.* 2014). Guiding research questions are: How will the resource efficient use of energy, water, materials and space change the future design of cities? Which 'loops' can be created within the city or parts of the city (e.g. neighbourhood, block, building)? Which developments (high tech and low tech) on the way to the loop city need be integrated into the city? How do people live and participate in regenerative cities?

The research approach of the ROOF WATER-FARM explores the impact of resource metabolism – with a focus on water management and food production – on infrastructures, architecture and urban form, public space, land use and density of the urban fabric, as well as city life. It follows the need to explore (also by informed design) subjects such as mixed use/multiple usage, low-tech and high-tech solutions, and public participation and education. The urban metabolism – thus the hypothesis of the project – leads to architectural and urban designs that rely on user engagement and new operational management models when it comes to implementation and management. Multifunctional infrastructures will have a spatial impact, as well as an impact on people's everyday use of water.

Research on everyday cultural water practice, blockages and resistance to urban-ecological and technological change (Heidenreich and Glasauer 1997, Bürgow 2014) support the research approach – that urban metabolism can only be achieved through a participatory approach. The development of multifunctional infrastructure can also alter people's views on how and where food could be produced in inner-city contexts.

The research case is the city of Berlin. The European and post-industrial city of Berlin reflects the contemporary Zeitgeist of local water, food and daily resource consumption. Different urban farming[4] or urban water initiatives[5] demand and develop integrated and decentralized infrastructure designs for basic needs, which necessarily include wellbeing or educational functions and consist of flexible low-tech and high-tech technologies. Aquacultural farming demands water-farming typologies, such as swimming gardens, fish ponds or types of water-farm greenhouses. These then become recognized as promising building-blocks and catalysts in water-sensitive cities integrating regenerative design/loop principles (Bürgow

2014). As traditional integrated farm systems (e.g. known from ancient Asian rice-paddy fish-farming or from the swimming gardens of the Aztecs in Mexico called *chinampas*), aquacultural typologies are newly valued as specific blue-green infrastructure within contemporary cities (Bürgow 2014). Their multifunctional design, due to combined water management and food production, can support the transformation process towards regenerative cities, such as loop cities. Aquaponic and hydroponic systems are available on the market 'for everyone from the advancing hobbyist to the beginning commercial grower' (PENTAIR Aquatic Eco-Systems Inc. 2015: 22).

In addition, the benefits of similar water-farm technologies in sustainable aquaculture have been broadly investigated, particularly within the transdisciplinary research field of ecological engineering (e.g. Guterstam and Todd 1990, Etnier and Guterstam 1991, Guterstam and Etnier 1996, Jana 2003, Bohemen 2005a, Steinfeld and Del Porto 2007). The combination of fish production, for example, with other agricultural livestock and crop production makes aquacultural farming more space, energy and resource efficient due to promising integration into urban space (Bürgow 2014). This type of farming reaches a higher productivity than contemporary practices of space-extensive organic farming due its use of fish manure for fertilization or nutrient-rich water from fish production for irrigation and fertilization (e.g. Hinge and Stewart 1991).

Early rooftop water-farm concepts[6] are associated with pioneering experiments in ecologically engineered aquaculture greenhouses in North America in the late 1960s/early 1970s, connecting with keywords such as Bioshelters, Living Machines or Solar Aquatic Systems (Todd and Todd 1984, Todd 1991, Guterson 1993, Todd and Todd 1993, Steinfeld and Del Porto 2004, Bohemen 2005b). Initial experimental research focused on the design and performance of aquatic ecosystem technologies for solar-based wastewater management within a tropical greenhouse mesocosm under colder climate conditions. The first European pilot water-farm greenhouse – Stensund Wastewater Aquaculture – combining decentralized wastewater management with aquaculture and hydroponic production modules was realized at a Folk College campus in Stensund, the Trosa archipelago community south of Stockholm, Sweden (Chan and Guterstam 1995, Guterstam 1996). Its novelty and uniqueness was the integration of the new water-farm infrastructure into a real-life community context and connecting it to the educational programme of the school (Bürgow 2014). Based on the case experiences applied, basic scientific research on aquaponic greenhouses showed the quantitative potential regarding nutrient recycling and space and energy efficiency (e.g. Rennert 1992, Junge-Berberovi *et al.* 1999, Staudenmann and Junge-Berberovi 2003, Graber and Junge-Berberovi 2008, Graber and Junge-Berberovi 2009), as well as market potential (e.g. Staudenmann and Junge-Berberovi 2003). A selection of benefits from water-farm technology targeting urban needs for sustainable transformation are summarized here (Bürgow 2014):

- Three-dimensional use of (water) space enables higher productivity per square metre.

- About 1kg of feedstuff is required to produce 1kg of high-quality fish (e.g. trout), whereas it requires 3kg of feedstuff for 1kg of poultry and 10kg for 1kg of beef or pork.
- A rule of thumb is that wastewater from 1kg fish can fertilize 5kg of vegetables, saving manure and giving a twofold use of water.
- There is no discharge of wastewater and, consequently, no land-based nutrient losses if the remaining fish manure is also recycled, e.g. via combined constructed wetland or productive soil systems.
- Arable land is used effectively. Greenhouse hydroponic systems can reduce agricultural land and water use by a factor of five to ten.
- Hydroponics produces eight to ten times more vegetable foodstuffs in the same area and time.
- Compared to soil-based plant production, hydroponic systems are suitable for lightweight rooftop applications.

In order to meet contemporary needs for decentralized and flexible integration of water and food technologies into existing urban spaces, ROOF WATER-FARM applies the benefits mentioned and goes a step further. It links established technologies for building-integrated water management and water reuse with aquaponic or hydroponic production of food on urban rooftops.

ROOF WATER-FARM modules

ROOF WATER-FARM, through its modular blue-green design, strives for mutual life support in the sense of a co-housing and co-working of man, fish and plants. The aim of applied research is to test the combination of new modular technologies for the safe treatment of building-related water flows (grey water, rainwater, black water) and its reuse for water-based food production via aquaponics or hydroponics.

At the technological level, ROOF WATER-FARM searches for alternatives to using drinking water and artificial fertilizer by addressing the high transport and energy expenditures of urban food production and supply with the development of local loop technologies. In brief, one can say that ROOF WATER-FARM combines water treatment (blue) with farm production (green).

Regarding the *blue*, two main water sources from buildings are the focus for local irrigation: grey water (from showers, sinks, kitchen, washing machines) and/or rain water (from rooftops), with black water (from toilets) being an additional source that is used as a basis for liquid fertilizer production for the local fertilization of plants with necessary nutrients and minerals, mainly nitrogen, phosphorous and potassium (NPK), which should meet the standards of a commercial fertilizer.

The quality aim for irrigation water is to meet European Union bathing water standards by applying the latest modular technologies, such as MBBR (Moving Bed Bio Reactor) for grey water or standard mechanical pre-treatment (filter systems) for rainwater. The results of the projects have shown that this is possible. A new membrane technology was tested for the unique approach to liquid fertilizer production out of black water. With this, ROOF WATER-FARM has contributed a technological module to target the shortage of phosphorous, which has been recognized as a much scarcer resource than fossil oil (e.g. Gerling and Wellmer 2005). Today it is mined to a great extent through socially and ecologically questionable methods, mostly in Africa (e.g. Schuh 2009).

Regarding the *green*, the ROOF WATER-FARM concept investigates two main water-farming types: aquaponics (fish and plants) and hydroponics (plants). Depending on the structural preconditions of the building and the users' needs, these water-farm types are lightweight compared to soil-based farming systems and, therefore, more suitable to apply to the surfaces of buildings (roofs, facades).

When combining the *blue* and the *green*, ROOF WATER-FARM investigates four ROOF WATER-FARM variants (Figure 9.1). It explores two optional water sources for irrigation combined with the two water-farming types, along with the appropriate process technology. The variants can be adapted according to the specific building typologies and their water production (e.g. patterns of daily water use: is there grey water production or not?) as well as the needs of the users (e.g. which products are liked?).

ROOF WATER-FARM variants focusing on aquaponics (fish and plants) are: variant I: reuse of grey water via treatment and transformation into process water (EU bathing water quality) for the cultivation of fish and building-related reuse (toilet flushing); the plants are fertilized via the nutrient-rich water from fish production; and variant III: usage of rainwater from the roofs for fish production; the plants are fertilized via the nutrient-rich water from fish production. ROOF WATER-FARMING variants focusing on hyprodponics (plants) are: variant II: reuse of grey water via treatment and transformation into process water (EU bathing water quality) for the cultivation of plants and building-related reuse (toilet flushing); safe usage of domestic black water and transformation into a liquid fertilizer; the fertilization of plants is adjusted via the crop-specific mix of the fertilizer with the process water; and variant IV: usage of rainwater from roofs for the cultivation of plants; safe usage of domestic black water and transformation into a liquid fertilizer; the fertilization of plants is adjusted via the crop-specific mix of the fertilizer with the process water.

The first ROOF WATER-FARM pilot plant was realized in Berlin-Kreuzberg and docked on to the existing decentralized water management infrastructure of Block 6 in the spring of 2014 (Figure 9.2). The integrated water concept at Block 6 in Berlin-Kreuzberg was originally developed as a project of the International Building Exhibition 1987, further researched as a model project of the Experimental Housing and Urban Development Program financed by federal and state funds until 1993, and finally, optimized and redesigned in 2006/07. Since the project's start, domestic wastewater produced by the tenants of Block 6 has been separated into the two flows: grey water and black water. The comparatively low-nutrient grey water from bathtubs, showers, washbasins, washing machines and kitchens is separated via a second water supply pipe. Since 2006/07, the grey water of around 250 tenants has been treated mechanically and biologically towards EU bathing water quality in a separate water-processing house and reused to flush

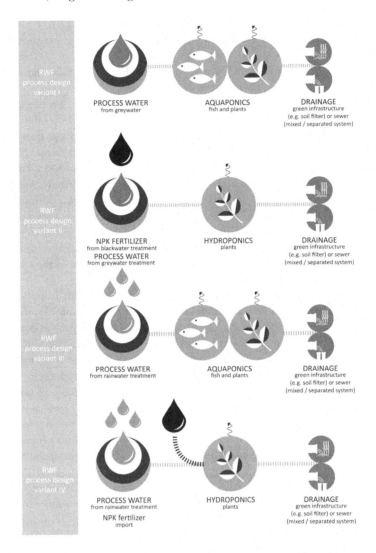

Figure 9.1 ROOF WATER-FARM variants combining building-integrated water reuse and water farming investigated in ROOF WATER-FARM

Source: TUB-ISR © ROOF WATER-FARM.

toilets and irrigate tenants' gardens. The project ROOF WATER-FARM extends the existing water reuse concept and uses the treated grey water (bathing water quality) for the production of fish and plants in the new greenhouse, while black water (from the toilets) is transformed into a liquid fertilizer. A water-farm greenhouse is constructed next to the water house, which contains grey water treatment technology and the newly constructed black water fertilizer plant. In the project, the hygienic quality of the irrigation water from the treated grey water and the

Figure 9.2 ROOF WATER-FARM pilot site with the greenhouse attached to the water
house situated in a courtyard of a housing block (Block 6) close to Potsdamer
Platz in Berlin

Source: Angela Million © ROOF WATER-FARM.

liquid fertilizer is monitored, and the fish and plants are also tested for relevant
micro-pollutants. The ROOF WATER-FARM pilot plant delivers further
process-technical data, such as for cost-benefit and lifecycle analyses or for investigating product quality and productivity of the technology.

From technology to space

The urban context calls for flexible and architecturally integrated applications,
which, on the one hand, increase the quality of life and create new livable urban
spaces and, on the other hand, need to provide infrastructure services for different
sectors, exploiting space in a multifunctional and participatory way. The research
on the diffusion of ROOF WATER-FARM technologies into the urban texture
focuses on three spatial levels: the city of Berlin, city quarters and neighbourhoods,
and individual building sites.

City of Berlin

The research focuses on identifying theoretical potential for the application of
ROOF WATER-FARM concepts on available rooftops in Berlin and is based on
previous research into building-integrated farming in Berlin, e.g. Zero Acreage

Farming Z-Farm[7] and studies in other Western cities, such as New York City (Ackerman *et al.* 2012). A geographical information system (GIS) is used to integrate parameters, such as rooftop dimensions, inhabitants, (waste)water flows, resource metabolism and building typologies (according to uses and social life). The datasets are generated primarily from geo-data and statistics from the City of Berlin.[8] When merging these datasets with basic figures on material flows of the ROOF WATER-FARM systems, citywide application potential according to the characteristics of the four different variants/concepts can be derived. The map of potential ROOF WATER-FARM applications focusing on flat rooftops in Berlin shows and visualizes the distribution of the innovative technologies in the urban realm. Due to the building-specific availability of the resources required, the four variants have different characteristics in their spatial diffusion potential, e.g. aquaponics with treated grey water (Figure 9.3).

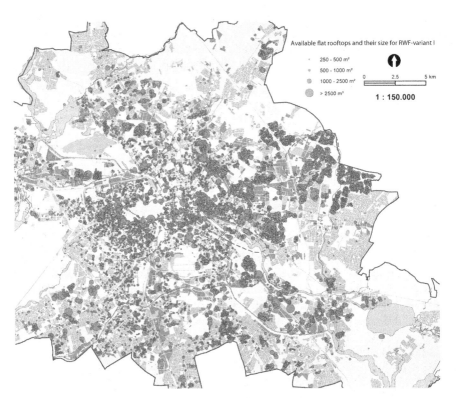

Figure 9.3 Spatial diffusion potential of ROOF WATER-FARM variant I on the flat roofs of the city of Berlin

Source: ROOF WATER-FARM, Database: Senate of Berlin, Department for Urban Development and Environment and Statistical Office for Berlin-Brandenburg.

The results of the citywide analysis also exhibit the diversity of building typologies within the 900km² city boundaries of Berlin. Only 13 per cent of the 536,000 individual buildings included have flat roofs larger than 50m² and therefore embody a theoretical usage potential. However, the analysis also showed that these potential buildings, which represent 45 per cent (4423ha) of the total building area are housing 57 per cent of the 3.4 million Berlin inhabitants. Sixty-five per cent of 'potential' buildings correspond to the 'residential' building typology, derived from the automated cadastral register of Berlin.

Analyzing the four ROOF WATER-FARM variants reveals that those variants using rainwater instead of grey water have larger overall area potential in Berlin. This is due to the fact that buildings with few or no inhabitants (hence, low wastewater production), such as factories, offices or supermarkets, are also suitable. Applying a conservative production footprint of 25kg of vegetables per square metre and 15kg of fish per 1m³ tank (surface area 1.2m²) and year, the production potential can be estimated throughout Berlin as either up to 165,000 tons of vegetables and 14,000 tons of fish per year with grey water use;[9] or up to 299,000 tons of vegetables and 10,000 tons of fish per year with rainwater use.[10]

Regarding the multifunctionality of ROOF WATER-FARM, potential consequences for the water sector have been outlined. In addition to savings in irrigation water, fertilizer, cooling and transportation in comparison to conventional production elsewhere, the potential impacts on the urban water sector have been evaluated. Therefore, the assumption that excess purified grey water may be used for flushing toilets in buildings with ROOF WATER-FARM has been included in the GIS simulation. Thereupon, the model estimated theoretical savings of 16 per cent of total domestic drinking water consumption in Berlin and 23 per cent less domestic wastewater to be treated in municipal wastewater treatment plants. Furthermore, collection and use of rainwater by ROOF WATER-FARM has the potential to tap up to 14 per cent of the total paved area in the city, which could be a significant contribution and relief for the central storm water infrastructure. On the downside, aquaculture might add nutrients to the citywide wastewater stream, and reduced water flow in the public water system might make a larger restructuring process necessary. Due to the complex and spatially diverse characteristics of the municipal sewage system, the impacts of the modified supply and drainage regime by ROOF WATER-FARM technologies have to be evaluated on a local scale.

City quarters and neighbourhoods

Case study areas within the city of Berlin were chosen by criteria such as:

- Location and dynamics of development, e.g. density, migration balance, new construction, demand for redevelopment;
- Building form and use, mapping of the building prototypes;
- Social setting, e.g. income, property ownership and housing rent, social services offered.

According to these criteria, three model areas were selected that serve as potential diffusion areas of the ROOF WATER-FARM technology:

1   Inner-city Berlin (Potsdamer Platz, Mierendorff-Insel);
2   Areas of transformation, inner void (Spreestadt, Rummelsburger Bucht);
3   Suburbs and border areas (Marzahn-Hellersdorf).

These model areas are interpreted as prototypes for distinctive dynamics, stakeholder networks and significant urban forms that are common within European cities. As part of the suburb model area, the large housing estate of Marzahn-Hellersdorf was selected for a quantitative and qualitative study of the roof surface potential as well as stakeholder networks to implement and manage ROOF WATER-FARMs.

For the residents and local actors such as local businesses and institutions, potential material and product resource cycles were analyzed and interwoven towards a future scenario plan to explore and visualize questions such as: who could operate and maintain a rooftop greenhouse, and who could distribute and consume the products? Are there non-commercial management strategies and who could be interested in them? How can greenhouse production and product consumption be organized locally, in the block, in the neighborhood, in the city district and beyond? What logistics are required and where can they be accommodated?

A network master plan graphically captures and displaces diverse stakeholder relations and material and product flows (Figure 9.4). Here, the actors are geographically localized and described by building type, land use and traffic infrastructure, and described so also by physical distances[11] between producer and purchaser. Different scenarios can be put into the network master plan: long-term business linkages in terms of a vision for a neighbourhood or city quarter, as well as activities that can be implemented in the short run. The network plan is therefore an analytical, conceptual working document that can be changed, for example as new actors enter the scene or actors drop out. It is also a communication tool, mainly due to the addition of street views and building perspectives, which test and depict the built integration of blue-green infrastructure in the urban landscape.

As part of the scenario and network master plan for Marzahn-Hellersdorf, for example, the cooperation between residents and a housing association was simulated: roof greenhouses can be a resident initiative and/or offered as a service and amenity by the housing association to the tenants. Due to the roof potential of the prefabricated housing estates, private investors may be interested in running a roof farm. The International Horticultural Exposition planned for 2017 in Marzahn could promote such activities. Daycare centers, cafeterias of schools and local businesses would be the buyers of local food products. On the roofs of these institutions public and educational spaces for food production could be implemented to serve as demonstration gardens for urban metabolism. Green and blue infrastructure could be tactile to provide a sensual experience.

The described site-specific actors and urban structures are visualized in the network master plan of the district Marzahn-Hellersdorf (Figure 9.5) and supple-

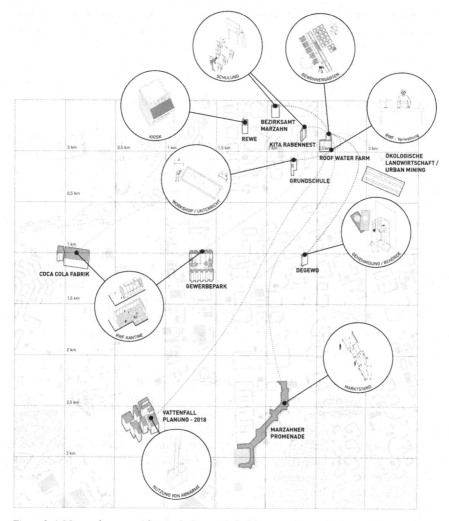

Figure 9.4 Network master plan including stakeholder network and resource cycles for the neighbourhood Marzahn-Hellersdorf

Source: TUB-ISR © ROOF WATER-FARM.

mented by narrative scenarios. This includes reflections on the accessibility of roof gardens and participation in urban space production. Finally exemplary user profiles and street views illustrate the everyday dimension of cross-sectorial management and the networking of material flows in the urban space of Marzahn-Hellersdorf.

Although network master plans will be different for each city quarter, based on our studies the implementation scenarios show the importance of the district level for food production, marketing and consuming, while for rainwater and wastewater management, the building and block level needs to be taken into consideration. Building-related service water and rainwater flows become the resource for regional production and marketing industries and their consumers.

Building sites: typologies, architectural site-layouts and ROOF WATER-FARM designs

At the building and site scales, the transferability of the ROOF WATER-FARM concept can be be proven via a process-technical, architectural and operational up-scaling for different typologies (residential, educational, commercial, accommodation/hotel, social and cultural, including industrial transformation buildings). Thereby, the different ROOF WATER-FARM variants are applied to the individual building typology responding to the specific resource metabolism of the building, i.e. the building's users (e.g. tenants, workers, students, etc.).

A residential building, for example, could apply rainwater or grey water reuse (suitable for all ROOF WATER-FARM variants I to IV) and the ROOF WATER-FARM could provide partial self-sufficiency with fresh fish and vegetables for the tenants. A school building might have an application for rainwater use (ROOF WATER-FARM variants III and IV), as usually not enough grey water is produced, whereby it could strive for partial self-sufficiency of the school canteen as well as becoming an educational farm/green classroom for the ROOF WATER-FARM concept (Figure 9.5).

In order to investigate the basic feasibility and adaptability of the ROOF WATER-FARM approach as a building-integrated concept, an integrated building study was performed focusing on typology 1 – residential. As an innovative spatial design research format, the building study focused on the transfer of the ongoing technology research up to the scale of a building block and its daily life metabolism. Thereby, questions of architectural site layout regarding fixed and fluid features (e.g. structural design of the building, daily resource flows by the tenants) were addressed as well as investment costs, costs-benefit calculations and operational aspects.

The residential building type, as the most common urban typology and suitable for all ROOF WATER-FARM variants, is of great importance, particularly regarding its everyday meaning in the city. This is also a result of the GIS working model investigating the existing flat-roof potential for the city of Berlin. It simulates the maximum possible transferability and shows that with a total number of 535,920 buildings, only 13 per cent (27,252) are flat-roofed buildings with more than 50m^2 usable area (RWF 2014). However, with 57 per cent, the majority of the 3.4 million recorded people in Berlin resides in these buildings.

Therefore, the basic ROOF WATER-FARM building study investigated two residential building entrances with five floors near the inner-city ROOF WATER-FARM pilot test site of Block 6 (Figure 9.6). Currently about 70 people live in the flat-roofed building at Dessauer Straße, 13–14 with an approximate gross floor area of 450m^2. All of the four ROOF WATER-FARM variants were simulated, which

Figure 9.5 ROOF WATER-FARM variants and their application to buildings
Source: TUB-ISR © ROOF WATER-FARM.

included water (the use of service water via decentralized rainwater or grey water treatment) and farming (both water-farm type options of aquaponics and hydroponics). In a first approximation, both the technical feasibility and the local coverage of water and nutrient demands for both water-farm types and the local coverage with fresh food for the tenants were estimated.

In summary, the following intermediate insights from both the pilot technology test site after the first year of operation and the first typological building study were derived for the residential building typology:

- Overall feasibility: technical feasibility appears secure. On similar structural designs, the transferability to other structures such as block, units, rows, etc. appears possible in principle, but needs a separate test along with implementation planning.
- Water-sensitive architectural design: for new building projects, the ROOF WATER-FARM consortium recommends the installation of a dual pipe

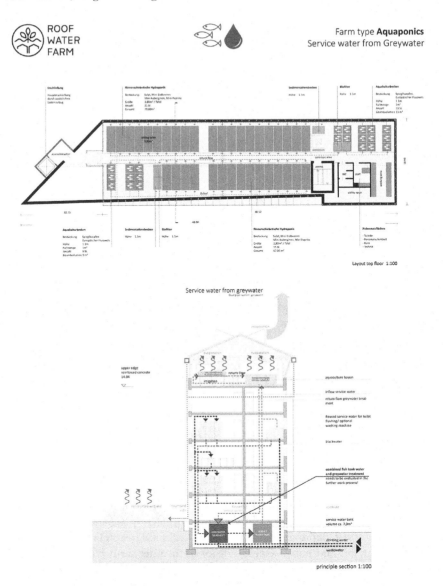

Figure 9.6 ROOF WATER-FARM residential building typology: Section and layout plan, variant 1 – aquaponics with grey water reuse

Source: Architekturbüro Freiwald und TUB-ISR © ROOF WATER-FARM.

system for the separation of grey water and potential reuse as household service water, complying with the standards of the building code. The total system costs, including the dual pipe network, are less than 0.6 per cent of the total apartment costs (Nolde 2014).

- Water-sensitive urban design: the ROOF WATER-FARM residential building type contains a promising transferability potential, particularly for new urban building projects or reconstruction, such as partial refurbishments. The ROOF WATER-FARM variants using grey water are thereby promising due to their high savings of drinking water, which also saves energy as the transportation of drinking water is reduced.

- Food-productive urban design: ROOF WATER-FARM products such as carp species, salad and strawberries showed good performance in the first ROOF WATER-FARM season, whereas for example peppers were more problematic. If opting to meet the local fresh food demands of the people living under the roof, the type of products grown can vary. The percentage of the building-integrated food supply thereby strongly depends on the available rooftop space and the number of floors in the building (the higher the building, the more people need to be supplied).

- Greenhouse water needs: the demands for irrigation and process water for the water-farm types can easily and securely be met if grey water is used. Grey water contributes 70 per cent of total wastewater in private households: assuming 100 litres of total wastewater, 70 litres are grey and 30 litres are black. If using rainwater from roofs (e.g. for aquaponics), theoretical calculations show that additional treatment of tank flushing water from fish production is often necessary; otherwise the water demand cannot be met for the total rooftop greenhouse space. Therefore, to integrate an additional module for the treatment and reuse of wastewater from fish tank flushing is an option to better meet water demands or to collect additional rainwater sources, e.g. from nearby rooftops.

- Greenhouse energy needs: in temperate climate zones, such as the city of Berlin, the demands, particularly in the colder seasons, are critical and need to be investigated according to the individual project site, since it contributes to the overall feasibility of the system. Synergistic potentials, such as through using excess heat from the building or technical infrastructure in the local neighbourhood, should be examined. Further potentials for optimization are in the area of greenhouse design (e.g. choice of material, design of the size and shape).

- Construction costs: it can be expected that the choice of larger rooftops reduces the proportional costs for the development. Alongside the planned development of exemplary commercial and non-commercial operation models for selected ROOF WATER-FARM building typologies some cost-benefit examples can be shown to affect value creation at the urban and neighbourhood scales.

The prototypical method is now used as a standard template for other building typological studies, particularly the development of ROOF WATER-FARM building passports. As a more generalized communication format, the building passports address major aspects of the transferability and flexibility of the ROOF WATER-FARM concept. They contain information cards providing specific features of the ROOF WATER-FARM typologies regarding:

1   The building (e.g. construction form, architectural layout, site information);
2   The building's water infrastructure (e.g. potentials and modules of grey water, black water, rainwater);
3   The prospective farm operation (e.g. exemplary water-farm designs, product ranges and potential harvests). This is intended to be another ROOF WATER-FARM information tool, which can be used to develop network plans at the urban and neighbourhood scales in creative dialogue and playful interaction with local actors.

Public relations, participation and education

The potential implementation of the ROOF WATER-FARM concept not only requires functioning technologies, but also needs to consider the acceptance and requirements of the potential operators, users and consumers (e.g. urban developers, investors, building owners, city administration, urban water works, urban farmers, residents). It has been shown within the research project that it is fruitful to communicate the applied research in a transparent and continuous manner from the beginning – e.g. via website, social media,[12] guided tours and events on the pilot site.

Aiming for a diversity of target groups and research process-related communication, the research team is working with different formats for user participation and acceptance of the technology within the urban population. Interviews and different user profiles (video/audio), combined with a web and communication campaign, help to involve a diversity of potential actors in the research and communication process. With the pilot site, the research team has a tangible small-scale example of the technology that exemplifies different aspects, e.g. hygiene questions, harvest potentials, noise, smell and taste, questions of design and materials, monitoring and maintenance and costs. Events, guided tours and hands-on workshops involve potential users directly in the research process and support the team with target-group specific questions and needs.

A workshop with students of a primary school (Figure 9.7) was held and has become a Marzahner fieldwork site for the development of communication and training materials for education and training providers. Information formats for users and decision-makers from the fields of urban and infrastructure development will be created in other research processes. Communication and education – for example, for tenants, investors and housing companies – are necessary and a prerequisite for acceptance of new technologies, which is one of the key interim results of the research network. The presentation of the shift of decision-making and responsibility levels (concerning user profiles, operator models, communication and education formats) is therefore integrated into the building and district studies. Potentials and risks of the new infrastructure will be analyzed on the levels of observation of the urban space, buildings and actors/users and prepared for communication and education formats.

The first results of this communication process and participatory approach are evident. The number of tours increases monthly, and the team guides a broad international circle of disciplines through the pilot site (Figure 9.8), the project's

Figure 9.7 Results of a ROOF WATER-FARM hands-on workshop in the Peter Pan Elementary School in Marzahn

Source: Diego Rigamonti © ROOF WATER-FARM.

Figure 9.8 Press tour of the German Academic Exchange Service (DAAD) as part of the National Year of Science 'Future Cities'

Source: Marius Schwarz © DAAD.

vision and its working steps. Different international media have reported on the research. The focus of most media reports was to present the ROOF WATER-FARM idea as a vision and perspective of a 'Future Citiy'. ROOF WATER-FARM technologies, combining water treatment and food production, are discussed as an example for a more resource-efficient, resilient urban texture. Through the pilot site, workshops, events, guided tours and press reports, a powerful potential for urban transformation starts to unfold: wastewater treatment – once just portrayed as a crisis and perceived as negative for cities' daily life now gets headlines in public news like 'wastewater farming', 'gold-water farming', 'going off the water grid', 'fruits and fish from roof in Berlin-Kreuzberg' and 'from the dishwasher to the homely vegetable bed'.[13]

With the web campaign and the process-related communication tools, the research team offers appealing elements to engage and enable a broad field of actors into the process of infrastructural development and urban design. Questions from workshops with experts (planning, design, architecture) and non-experts (schools, citizens) became integrated into the research process. Techniques of network-mapping and narrative scenarios will evolve into practical guidelines and user-related learning elements for different target groups (design/planning, education and everyday life). Innovations research and a participatory multicriteria analysis of relevant factors for implementation and operation of ROOF WATER-FARM concepts facilitate the process. Key motivations, drivers and obstacles for the realization of the cross-sectorial infrastructure are being extracted from interviews with diverse experts, practitioners and authorities. After mapping and evaluating the identified elements, stakeholder-specific implementation and communication strategies can be developed. The practical guidelines will be published online as a comprehensive brochure, info-graphics and a toolkit for ROOF WATER-FARM users and developers.

Conclusion

ROOF WATER-FARM researchers looked at water and farming technology, architectural and urban design, neighbourhood development, public participation and empowerment. The development of individual and combined water treatment components enables site- and user-specific applications and shows possibilities for spatial transferability to different building types as well as to different neighbourhoods. Three spatial research scales (city, neighbourhood, building site) are analyzed to clarify the potential of the diffusion of the technology into the urban realm and to visualize that potential as different building, district and actor-related innovations. Atmospheric changes/transformation within the urban texture, e.g. rooftops with greenhouses, infrastructures for harvesting, processing and storage, and marketing and consumption opportunities, will be visualized. The challenge of this research is to consider this concept from the scale of a single water pipe to the scale of a city.

Within the vision of regenerative cities and loop cities, the ROOF WATER-FARM project derived from explicitly focusing on water resource cycles and their interface with food production. If it develops further as part of food planning and production within the city, the ROOF WATER-FARM concept will need to take into closer consideration other resource cycles and infrastructure sectors (e.g. energy) and infrastructure services (e.g. leisure, education). If multifunctional infrastructures such as ROOF WATER-FARM become part of the city, it will increase the multifunctionality of urban space. Thereby, green open space will also include new infrastructural services and, consequently, also become revalued as green or blue-green infrastructure. This impact will be spatially seen, felt and lived by urban citizens.

Although the ROOF WATER-FARM research team expected widespread skepticism towards the developed technology, the project showed that there is interest and curiosity from the public towards the project, and it is strongly associated with positive urban visions for future cities. Accordingly, tools and formats for participation, education and enablement can directly work with the initial interest to foster further public acceptance and implementation of ROOF WATER-FARM technologies, designs and products.

Although the developed technology offers site- and user-specific adaptations, there are challenges in implementation. There are extra investment costs for water recycling, the pipe systems and farm equipment, or for safe access to roofs (elevator, etc.) that have to be made. In implementation, it is probably often the case that the investor (unless she or he is also a user of the building) is not the primary beneficiary (which would be e.g. tenants). New operating models and changing responsibilities from design to managing need to be developed and further tested. The case study of Marzahn shows that a cooperatively organized model or commercial business strategies offer options to build and run a ROOF WATER-FARM. Further non-commercial operation models illustrate the potential of multifunctional infrastructure as places of education for sustainability.

Acknowledgements

A special thanks goes to our research partners and their great effort to make this interdisciplinary research project happen. This includes Architekturbüro Freiwald who was part of the integrated building study. Thank you also to our student assistant researchers Felix Bentlin, Tim Nebert, Jürgen Höfler, Heiko Ruddigkeit and Svea Ruppert who translated scientific facts and figures into graphics and were part of design studies. Last but not least thank you to Diego Rigamonti for his great work in GIS and map analysis.

Notes

1   ROOF WATER-FARM partners are:
    • TU Berlin – Department for Urban Design and Urban Development at the Institute of Urban and Regional Planning (ISR); ZEWK kubus – Cooperation Center for Environmental Issues
    • Fraunhofer Institute UMSICHT, Oberhausen: www.umsicht.fraunhofer.de
    • Nolde & Partner, Berlin: www.nolde-partner.de
    • Terra Urbana Umlandentwicklungsgesellschaft mbh, Zossen: www.terraurbana.de
    • inter 3 – Institute for Resource Management GmbH, Berlin: www.inter3.de
    • Senate Administration for Urban Development and Protection of the Environment, Berlin.
2   Program website: www.bmbf.navam-inis.de
3   Project website: www.roofwaterfarm.com/en/
4   See http://prinzessinnengarten.net; http://contemporaryfoodlab.com; www.zfarm.de
5   See www.flussbad-berlin.de; http://berliner-wassertisch.net
6   See http://nysunworks.org/thesciencebarge; http://brightfarms.com; http://lufa.com; http://urbanfarmers.com
7   Project website: www.zfarm.de
8   Senate of Berlin, Department for Urban Development and Environment and Statistical Office for Berlin-Brandenburg.
9   If considering an annual requirement of fresh vegetables/fruit of 68.7kg per person (Gesellschaft für Konsumforschung 2009), 2.4 million people could be theoretically supplied. If considering an annual need of fresh fish of 5.2kg per person (Gesellschaft für Konsumforschung 2012), 2.7 million people could be theoretically supplied.
10 Consequently, 4.3 million people could be theoretically supplied with fresh vegetables/fruit or 1.9 million people with fresh fish.
11 Physical distance and travel time are still used in Germany to describe access to food and local supply (Uttke 2011).
12 www.facebook.com/roofwaterfarm
13 www.roofwaterfarm.com/en/category/medien-2/

References

Ackerman, K., Plunz, R., Conard, M., Katz, R., Dahlgren, E., Culligan, P. (2012): *The Potential for Urban Agriculture in New York City, Growing Capacity, Food Security, & Green Infrastructure*, Urban Design Lab, The Earth Institute, Columbia University.

AECOM (ed.) (2013): *Water Sensitive Urban Design in the UK.* Ideas for Built Environment Practitioners. London: CIRIA.

Barton, H., Grant, M., Guise, R. (2010): *Shaping Neighborhoods for Local Health and Global Sustainability.* 2nd Edition. London and New York: Routledge.

Beatley, T. (2000): *Green Urbanism: Learning from European Cities.* Island Press, Washington, DC.

BMBF (Federal Ministry of Education and Research) (2012): Funding priority Sustainable Water Management NaWaM within the framework programme on Research for Sustainability Development. Bonn: BMBF.

Bocek, A. (1996): Introduction to Polyculture of Fish: Water Harvesting and Aquaculture for Rural Development. [online]. Auburn University, Alabama, USA.

Bohemen, H. von (ed.) (2005a): *Ecological Engineering: Bridging between Ecology and Civil Engineering.* Boxtel: Aeneas Technical Publ.

Bohemen, H. von, (2005b) Green Machines (Living Machines) for Processing Wastewater. In: H. von Bohemen (ed.) *Ecological Engineering: Bridging between Ecology and Civil Engineering,* pp. 73–78. Boxtel: Aeneas Technical Publ.

Bürgow, G. (2014): Urban Aquaculture: Water-sensitive transformation of cityscapes via blue-green infrastructures. Dissertation 25.11.2013, Technische Universität Berlin. Schriftenreihe der Reiner-Lemoine-Stiftung, Herzogenrath: Shaker Verlag. ISBN 978-3-8440-3262-8. doi:10.2370/9783844032628. www.shaker.de/de/content/catalogue/index.asp?lang=de&ID=8&ISBN=978-3-8440-3262-8

Bürgow, G., Million, A., Steglich, A. (2015): Urbane (Ab-)Wasser- und Nahrungsmittelproduktion – Neue partizipative und multifunktionale Infrastrukturen in der Stadt. In: RaumPlanung 180/ 4-2015, S. 54-65, Themenheft: Grüne Infrastruktur in urbanen Räumen. Informationskreis für Raumplanung e. V. (IfR). Dortmund. www.ifr-ev.de/index.php?id=1057&type=98

Chan, A., Guterstam, B. (1995): Stensund Wastewater Aquaculture: A Small-Scale Model for Sustainable Water Resource Management in Urban Areas. In: J. Niemczynowicz (ed.) (1995) *Integrated Water Management in Urban Areas: Searching for New, Realistic Approaches with Respect to the Developing World,* pp. 315–328. Proceedings of the International Symposium, UNESCO-IHP, September 1995. Lund, Sweden.

Costa-Pierce, B., Baker, D., Desbonnet, A., Edwards, P. (eds) (2005): *Urban Aquaculture.* Oxford: CABI International.

DIFU (Deutsches Institut für Urbanistik) (2013): *Intelligente und multifunktionelle Infrastruktursysteme für eine zukunftsfähige Wasserversorgung und Abwasserentsorgung Vorstellung der Verbundprojekte: Ziele, Projektpartner und geplantes Vorgehen.* Bonn: BMBF.

Etnier, C., Guterstam, B. (eds) (1991): *Ecological Engineering for Wastewater Treatment: Proceedings of the International Conference at Stensund Folk College, Sweden, March 24-28.* 1st ed. Gothenburg, Sweden: Bokskogen.

European Commission (EC) (2010): *LIFE Building up Europe's Green Infrastructure: Addressing connectivity and enhancing Ecosystem functions.* Luxembourg. http://ec.europa.eu/environment/life/publications/lifepublications/lifefocus/documents/green_infra.pdf

Ferguson, F. (ed.) (2014): Make_Shift City. Die Neuverhandlung des Urbanen. Jovis Verlag. Berlin. ISBN 978-3-86859-223-8.

Gerling, J. P., Wellmer, F.-W. (2005): Wie lange gibt es noch Erdöl und Erdgas? *Chemie unserer Zeit* 39: 236–245.

Gesellschaft für Konsumforschung (Gfk) & Fischinformati-onszentrum e.V. (FIZ) (2012): www.ischinfo.de/index. php?1=1&page=infograiken

Gesellschaft für Konsumforschung (Gfk) & fruchtportal (fp) (2009) www.fruchtportal.de/aktuelles/lesen/23612/ Jeder-Bund.

Gorgolewski, M., Komisar, J., Nasr, J. (2011): *Carrot City: Creating Places for Urban Agriculture.* 1st ed. New York: Monacelli Press.

Graber, A., Junge-Berberovi , R. (2008): Wastewater-fed Aquaculture, Otelfingen, Switzerland: Influence of System Design and Operation Parameters on the Efficiency of Nutrient Incorporation into Plant Biomass. In: J. Vymazal (ed.) *Wastewater Treatment, Plant Dynamics and Management in Constructed and Natural Wetlands*, pp. 299–310. 1st ed. Netherlands: Springer Verlag.

Graber, A., Junge-Berberovi , R. (2009): Aquaponic Systems: Nutrient Recycling from Fish Wastewater by Vegetable Production. *Desalination* 246: 147–156.

Grabow, B., Uttke, A. (2010): Leitbilder Nachhaltiger Stadtentwicklung. Von der Lokalen Agenda zur Nachhaltigkeit als strategischem Rahmen. *PLANERIN* 6(10): 22–25.

Guterstam, B. (1996): Demonstrating ecological engineering for wastewater treatment in a Nordic climate using aquaculture principles in a greenhouse mesocosm. Ecological Engineering 6: 73–97.

Guterstam, B., Todd, J. (1990): Ecological engineering for wastewater treatment and its application in New England and Sweden. *Ambio* 3: 173–175.

Guterstam, B., Etnier, C. (1996): The Future of Ecological Engineering. In: J. Staudenmann, A. Schönborn, C. Etnier (eds) *Recycling the Resource: Proceedings of the Second International Conference on Ecological Engineering for Wastewater Treatment, School of Engineering Wädenswil – Zürich, September 18 – 22, 1995*, pp. 99–104. Zürich-Uetikon: Transtec Publications.

Guterson, M. (1993): Living Machines: Putting human waste back in its place: at the bottom of the food chain. *Context – A Quarterly of Humane Sustainable Culture* 35: 37.

Heidenreich, E., Glasauer, H. (1997): *Zur Wasser kultur in städtischen Privathaushalten und zu den Spielräumen ihrer Veränderung.* WasserKultur-Texte Band 27. http://nbn-resolving.de/urn:nbn:de:hebis:34-2008051921605

Hinge, J., Stewart, H. (1991): Solar Wastewater Treatment in Denmark: Demonstration Project at Danish Flokecenter for Renewable Energy. In: C. Etnier, and B. Guterstam (eds) *Ecological Engineering for Wastewater Treatment: Proceedings of the International Conference at Stensund Folk College, Sweden, March 24–28, 1991*, pp. 181–187. Gothenburg, Sweden: Bokskogen.

Hoyer, J., Dickhaut, W., Kronawitter, L., Weber, B. (2011): *Water Sensitive Urban Design: Principles and Inspiration for Sustainable Stormwater Management in the City of the Future.* Berlin: Jovis.

Jana, B. B. (ed.) (2003): *Sustainable Aquaculture: Global Perspectives.* Binghamton, NY: Food Products Press.

Junge-Berberovi , R., Staudenmann, J., Todt, D. (1999): Wastewater-fed Aquaculture in Temperate Climates: Optimizing the Recycling Rates via Primary Production. In: B. Kløve, C. Etnier, P. Jenssen, and T. M hlum (eds) *Managing the Wastewater Resource: Ecological Engineering for Wastewater Treatment*, Ås, Norway.

Lefebvre, H. (2016): Das Recht auf Stadt. (Le droit à la ville, Paris 1968/2009, übersetzt von Birgit Althaler). Hamburg: Edition Nautilus. ISBN 978-3-96054-006-9

Libbe, J., Beckmann, K. (2010): *Infrastruktur und Stadtentwicklung: Technische und soziale Infrastrukturen – Herausforderungen und Handlungsoptionen für Infrastruktur- und Stadtplanung.* Berlin: Deutsches Institut für Urbanistik.

Lyle, J. T. (1994): *Regenerative Design for Sustainable Development*. New York, NY: Wiley.

Million (née Uttke), A., Steglich, A., Bürgow, G. (2013/2014): Participatory Blue-Green Infrastructure, Master class of Urban Design and City and Regional Planning, Winter Semester 2013/2014, TU Berlin. [online] http://urbandesign.staedtebau.tu-berlin.de/lehre/ma-srp-ud, www.roofwaterfarm.com/en/news/

Nolde, E. (2014a): Water and Energy Recycling in a Residential Passive House. Poster www.nolde-partner.de/node/32/

Nolde, E. (2014b): Das ROOF WATER-FARM Projekt, Energie und Stoffkreisläufe dezentral mittels Abwasserrecycling schließen. *fbr-wasserspiegel* 3/14, 16–17.

Overmeyer, K., Frech, S., Knödler, L., Scheuvens, R., Steglich, A., Ratzenböck, V., Kopf, X. (2014): *Räume kreativer Nutzungen. Potenziale für Wien*. Verlag für Moderne Kunst, Nürnberg.

PENTAIR Auqatic Eco-Systems Inc. (2015): *2015 Master Catalog. Global Solutions for the future of Aquaculture*. PENTAIR Auqatic Eco-Systems Inc.

Rennert, B. (1992): Simple Recirculation Systems and the Possibility of Combined Fish and Vegetable Production. In: *Progress in Aquaculture: Proceedings of 4th German-Israeli Status Seminar, October 30-31, 1990*, GKSS-Forschungszentrum Geesthacht, Germany. Edited by EAS Spec. Public., pp. 91–97. EAS Spec. Public.

RWF (2014): Zwischenbericht des Projektkonsortiums Roof Water-Farm. www.bmbf.de/foerderungen/16719.php

Schuh, H. (2005): Ohne Phosphat läuft nichts. [online] www.zeit.de/2005/23/N-Phosphatkrise/komplettansicht (18.08.2014).

Staudenmann, J., Junge-Berberovi , R. (2003): Recycling nutrients from industrial wastewater by aquaculture systems in temperate climates (Switzerland). *Journal of Applied Aquaculture* 13: 67–103.

Steinfeld, C., Del Porto, D. (2004): Growing away wastewater: Constructed ecosystems that clean wastewater. *Onsite Water Treatment* 1: 44–53.

Steinfeld, C., Del Porto, D. (2007): *Reusing the Resource: Adventures in Ecological Wastewater Recycling*. Concord, Mass.: Ecowaters Books.

Stewart, H., Chan, G., Hinge, J. (1991): Energy Budget for a Polyculture-based Wastewater Treatment System. In: C. Etnier, and B. Guterstam (eds) *Ecological Engineering for Wastewater Treatment: Proceedings of the International Conference at Stensund Folk College, Sweden, March 24–28, 1991*, pp. 181–187. Gothenburg, Sweden: Bokskogen.

Todd, J., Todd, N. J. (1984): *Bioshelters, Ocean Arks, City Farming: Ecology as the Basis of Design*. San Francisco: Sierra Club Books.

Todd, N. J., Todd, J. (1993): *From Eco-cities to Living Machines: Principles of Ecological Design*. Berkeley, California: North Atlantic Books.

Uttke (2011): Old and Emerging Center. Local food markets as today's anchors in urban centers. *disP* 185: 2.

Verbücheln, M., Grabow, B., Uttke, A., Schwausch, M., Gassner, R. (2013): *Szenarien für eine integrierte Nachhaltigkeitspolitik – am Beispiel: Die nachhaltige Stadt 2030 Band 2: Teilbericht Kreislaufstadt 2030*. Dessau-Rosslau: Umweltbundesamt.

World Future Council (WFC) & Hafen City University (HCU) (2010) *Regenerative Cities. Commission on Cities and Climate Change*. www.worldfuturecouncil.org/ileadmin/user_upload/papers/WFC_Regenerative_Cities_web_final.pdf

10 Associating agriculture with education and recreation in Queens, New York

Steven Buchanan

Food as an abstract idea: ecological disconnect

Through the collection of elements and massive redundancies causing us to connect, food is an important part of culture. In early cities, ports and food markets shaped city planning and social structure. Equal and educated food transactions kept families close to their food source until the introduction of the railroads led to a national food network, which was no longer contained by geography. Past social events of buying and selling food became anonymous as participants became inevitably apart but symbiotically connected.

By externalizing cost and destruction, the notion that humans are separate beings not affected by or affecting earthly processes has lead to habits of thought and intuitions based on separation. The loss of environmental values in western culture is less because society is genuinely uncaring, but rather because western culture does not teach how to love, care, and respect the natural world.

Offering less choice amid the perception of abundance, the large-scale effort of marketing causes unnatural demand and purposely obscures society's relation between food and farming. Understanding the context of a meal, how it arrives on our plates and how it affects the ecosystem is necessary for the study and practice of sustainable agriculture.

Global impact of agriculture

Habitat destruction

Growing human populations and the unprecedented rate of urbanization are leading to increasing resource demands for agriculture. Habitat modification and destruction, driven largely by industrial agricultural demands, pose a direct threat to biodiversity around the globe. Published in the journal *World Futures*, the article "World Population, Food, Natural Resources, and Survival" revealed that 6 kilograms of plant protein fed to livestock provides 1 kilogram of meat protein (Pimentel, 2003). According the United Nations Educational, Scientific, and Cultural Organizations 2010 report "Energy Flow, Environment and Ethical Implications for Meat Production," the livestock population in the United States consumed more than seven times as much grain as consumed directly by the entire American population (UNESCO, 2010). Consequently at the turn of the millennium the United States food production system used about 50 percent of the total United States land area, 80 percent of the fresh water, and 17 percent of the fossil energy used in the country (Pimentel, 2003). At the global level, the Food and Agricultural Organization of the Unites Nations 2006 report "Livestock's Long Shadow: Environmental Issues and Options" stated that live-stock production accounts for 70 percent of all agricultural land and 30 percent of all land surfaces on the planet (Steinfeld *et al.*, 2006). Still the average American meat-based food system is necessitating increasing energy, land, and water resource demands.

Ecological disturbance

Scientists from Radboud University in Nijmegen and the Dutch Centre for Field Ornithology and Birdlife Netherlands compared long-term data sets for both farmland bird populations and chemical concentrations in surface water (Hallman *et al.*, 2014). Published in the journal *Nature*, the study "Declines in Insectivorous Birds are Associated with High Neonicotinoid Concentrations" found that in areas where water contained high concentrations of imidacloprid—a common neonicotinoid pesticide—bird populations tended to decline by an average of 3.5 percent annu-ally: relating a widely used class of insecticides, called neonictinoids, to population declines across 14 species of birds (Hallman *et al.*, 2014). The elimination of certain weeds that harbor certain insects may be resulting in reduced survival, growth, reproductive rates, and amounts of healthy habitats for insectivorous bird populations as well. Diminishing bird populations, possibly including beneficial natural enemies, are increasing agricultural reliance on artificial inputs to control significant pest outbreaks to prevent yield loss in crops.

Published in the British Ecological Society's journal *Functional Ecology*, the study "Chronic Impairment of Bumblebee Natural Foraging Behaviour induced by Sublethal Pesticide Exposure" used radio frequency identification tags to monitor the day-to-day behavior of bumblebees (Gill and Raine, 2014). They found that prolonged pesticide exposure significantly affects individual bee behavior, including pollen collection and the choice of flowers (Gill and Raine, 2014). The study shows that this could have consequences for the growth and survival of colonies, further affecting the associated agricultural output. Current agricultural practices are disrupting ecosystems, resulting in increased artificial input to maintain expected yield outputs.

Antibiotic resistance

Once used sparingly to preserve the effectiveness to treat serious bodily infections caused by bacteria, antibiotics' two most common jobs include first to help live-stock, poultry and fish on industrial farms grow bigger and faster and second to help the animals avoid disease in unsanitary conditions in which they are raised

(Emanuele, 2010). According to a report by the Food and Drug Administration, approximately 80 percent of all antibiotics used in the United States are fed to farm animals (Emanuele, 2010). Respectively, bacteria continue to evolve more elaborate defenses that antibiotics can no longer cure. A study conducted by researchers at Johns Hopkins University examined flies near broiler poultry operations and found that many of these flies living near these operations carried antibiotic resistant enterococci and staphylococci (Evans *et al.*, 2009). Excessive antibiotic consumption has finally led us to the current epidemic of patients for whom there is no therapy. Increased resistance to antibiotics in an industry with few productive breeds leaves the industry open to infections that could wipe out a majority of livestock.

Livestock living conditions

Cattle, sheep, and pigs have frequently been seen as unintelligent animals that have lost their adaptations and cognitive abilities in the course of domestication. Consequently, livestock on industrial farms are commonly raised in environments that restrict the animals from expressing their physiological distinctiveness. Resent research is now assessing the cognitive ability in domestic animals including cattle, pigs, sheep, and dogs. Published in the journal *Animal Behavior*, the paper "Pigs learn what a mirror image represents and use it to obtain information" by researchers at the University of Cambridge revealed that pigs remembered how their own movements appeared in the mirror, and were able to apply that knowledge to a separate situation involving a hidden food bowl (Broom *et al.*, 2009). Pigs are one of only a few known species capable of assessment awareness in which they can understand the significance of a situation in relation to themselves, over a short period of time.

As for the cognitive abilities of cows, Katherine A. Houpt's book, *Domestic Animal Behavior for Veterinarians and Animal Scientists* states that cows can learn to distinguish between cattle and are less stressed in a frightening situation when familiar cattle are present (Houpt, 1982).

Rosamund Young of Kite's Nest Farm in Worcestershire, England, has observed, documented, and interpreted her experiences looking after the family farm's herd in her book *The Secret Life of Cows* (Young, 2003). Contrary to many modern farming practices that leave no room for displays of natural behavior, practices of Kite's Nest Farm involve no forced weaning and no separation of young from siblings or mother (Young, 2003). The most powerful relationship for a cow is that between a mother and baby. If given the opportunity, a mother cow may nurse her calf for as long as three years with the bond continuing after weaning; mothers and their children remain close to each other for life (Young, 2003). Adversely, as in many modern farming practices, baby calves are not allowed to nurse because cow's milk is intended for humans. In the book *The Ethics of What We Eat: Why Our Food Choices Matter* (Singer and Mason, 2007), John Avizienius, a senior scientific officer at the Farm Animal Department at the Royal Society for the Prevention of Cruelty to Animals in Britain, discusses what happens when calves are taken from their mothers as soon as two hours after birth. Once separated from their calves, the mothers cry and bellow for hours, even days (Singer and Mason, 2007). Just as the calves cry for their mother, the mother cows will search for their babies, visibly distressed, for a period of at least six weeks (Singer and Mason, 2007). The strong social bond between mother and calf explains the strong social bonds via the formation of "cow cliques" spending time together during daily activities. The conditions in which cattle are raised evoke strong emotions such as pain, fear, and even anxiety, yet if provided the proper conditions feelings of happiness can also be evoked.

Unnatural selection

Increased reliance on biotechnology companies deters the use and reproduction of heritage breeds in place of high-producing breeds dependent on commercial feeds, antibiotics, and other inputs of industrial agriculture. The Food and Agriculture Organization's 2007 publication, *State of the World's Animal Genetic Resources* reveals that 190 of the 7,600 breeds recognized by the Food and Drug Administration's Global Databank for Farm Animal Genetic Resources have become extinct in the past 15 years while a further 1,500 are considered "at risk" of extinction (FAO, 2007). Many of these heritage breeds provide important benefits and have many valuable characteristics such as disease resistance and extreme climate tolerance, and their bodies can be better suited to living on a pasture. For example, during a recent drought, a few of Uganda's farmers who had not traded their native Ankole cattle were able to move their herd to a far away water source. The farmers who replaced their native Ankole cattle with the Holstein-Friesians lost their entire herds (FAO, 2007). Scientists of the Food and Agriculture Organization predict that Uganda's native Ankole cattle could go extinct in the next 20 years because the Holstein-Friesians rapidly replace them. The return to more diverse networks and less concentrated industry will strengthen humanity's ability to provide food through climate change or disease stricken times.

Global impact

Industrial agriculture, sustained by the common western diet, are affecting many major categories of environmental damage including deforestation, erosion, fresh water scarcity, air and water pollution, climate change, biodiversity loss, social injustice, the destabilization of communities, and the spread of disease. Our choices no longer have only a local impact. An understanding of how consumption affects global biodiversity, ecosystem and environmental health, and human and animal health needs to be relevant in our decisions about food. As a collective whole, we must end the continuation of our current status quo and begin to listen to the needs and desires of all species.

The sacred in the city

Urban agriculture

During technologically mediated times, the postmodern experience offers little access to nature often making ideas the primary model of delivery for connecting to nature. Actual experiences must also connect humans to nature. Cities should not be isolated from the agricultural processes that sustain agricultural needs; they should be integrating these natural processes into their built environments. The most desolate urban landscapes can be made into fertile ground. From rooftops and vacant lots to vertical farms and edible landscapes, urban food production recognizes existing habitats, protects them, and benefits them by managing the land. During unprecedented times of urbanization, innovative urban agriculture can require little time, transportation resources, and money for the production of fresh and nutritious food.

Changing the way Americans eat begins with experiencing first-hand a humane and sustainable way of practicing agriculture. Pasture-based husbandry and vocational school in urban environments are an opportunity to teach city-dwellers about small-scale food production in contrast with large-scale industrial techniques. If done properly, chicken, rabbit, and bee keeping are perfectly safe practices in urban environments and can even restore the land. Otherwise useless vegetable scraps can be fed to chickens in return for farm-fresh eggs, and city honey is an edible record of available nectar in the urban landscape. Urban farming for recreation helps build a visceral understanding of how ecology works and how the plants and animals involved are affected.

Policy

In the book *Eat the City*, Robin Shulman describes how the combined efforts of transformative politics, urban planning, and public implementation can bring about a public spiritedness around local growing. In 1978, the Department of General Services, the agency in charge of all city-owned property, started Operation GreenThumb. Frank Silano was hired to manage the program and developed the ambition of renting out 10,000 lots to gardeners for a dollar per year per lot, furthering the process of turning green a portion of the most ravaged land in urban America. Some 21 agencies offered horticultural advice, seeds, topsoil, and tools. New York City entered the business of manufacturing soil using, leaves, tree trimmings, and grass cuttings to produce 200 tons of topsoil a month. By 1994, the project now called the Urban Gardening Program had expanded to 23 cities and was producing US$16 million worth of food each year (Shulman, 2012). Large-scale urban agricultural efforts assist in the restoration and development of more self-sustaining cities.

Through the integration of urban agriculture and participative strategies into policy and city planning, we can develop a framework for food justice by creating spaces and incentives for people to meet over food and agriculture. Change that is place-based can reclaim neighborhoods from blight, build communities, and reform and redeem cities. First the public must be given the right tools, knowledge, and resources necessary to establish and maintain urban agriculture.

Place-based implementation

This architecture project completed in the under-graduate program at Alfred State College, a state university of New York, explores the social and spatial aspects of an Agricultural Demonstration and Education Facility that exemplifies an alternative model of agriculture and radically rethinks current western agricultural means and methods. Prominently visible from many parts of New York City (including the United Nations Headquarters and the Empire State Building) and easily accessible through the East River Ferry, the facility sits on 10 acre site at the end of Hunters Point in Queens, New York. In order to reduce the impact of humankind on diversity, this facility will offer insight and resources for the public to address issues within our food system.

Comprehending the history of industrial agriculture and its increasing effects on the environment is crucial to the rethinking of current agricultural practices. The first component of this transformative educational experience is a **museum** whose three upward-rising glazed shards respectively symbolize the effects of pesticide usage, the conditions in which livestock are raised, and the profit-driven control in western agricultural models. Once food is related with farming and with the land,

Figure 10.1 Interior views of the museum

Source: Produced by author.

Figure 10.2 Ariel view of Hunters Point
Source: Produced by author.

fully informed food-related transactions will reconnect both the eater and the eaten to the biological reality of agriculture. Mother Earth is showing us signs of imbalance and is asking us human beings to respond and act on our sacred role as guardians of this planet.

Components of this rethinking include a **farm** that demonstrates the inter-workings of a humane, sustainable, and organic way of practicing agriculture. Along with composting, rainwater capture and crop rotation, students participate in the planting, watering, weeding, and harvesting of a wide variety of fruits and vegetables. Growing organic produce by a range of methods will demonstrate key measures that can sustainably establish and maintain local agriculture in an urban environment. Hydroponic window and roof-top farming, aquaponic farming, edible green walls, and raised-bed gardening demonstrations will enable the public to take action and lessen their dependence on industrial agriculture. Animals are an integral part of this urban farm. Through the expression of their physiological distinctiveness, heritage cattle, hogs, and laying hens perform ecological services for another. Through activities as simple as daily hand feeding, visitors can develop a better sense of an individual animal's health and appetite.

Figure 10.3 View from farm
Source: Produced by author.

An **education facility** will further help visitors understand the ecological processes on which agriculture depends. Visitors learn about soil fertility, photosynthesis, pollination, and botany, as well as nutrition and the history of farming in New York City. The program demonstrates how choices about food affect the health of humans, communities, and the environment. The combination of both ideas and actual experiences build a multi-dimensional relationship to nature. Its **event space** will offer lectures and meetings addressing public programming and local organizational efforts regarding social issues within current agricultural models: a space for the public to hear about what's going on in their community, to join forces with other supporters, and to get an idea out there. The facilities **store** will offer the tools, knowledge, and resources to assist in living a sustainable life. Composting bins, planting kits, instructional packages, and educational information will encourage and enable local agricultural efforts. Additionally, a farm-to-table **restaurant** will celebrate the food produced on the site by serving it in a way that retains its fresh, authentic flavors. A demonstration kitchen with cooking classes and educational events offer to both adults and kids the full benefit of collective culinary learning. Volume choices can minimize waste and the local foods menu will give the public choice.

A **composting facility**, acting as a heat source for a greenhouse located directly above, turns waste into resource offering to the public an alternative way of viewing garbage as a measure of who we are, rather than yet another difficulty. Once society understands how to wisely consume, compost, and recycle in order to ensure a sustainable future, it will be able to create a sense of purpose around the importance of the environment, local food, and soils.

A **farmers' market** will provide a point of distribution for sustainable farming efforts that are being implemented in the Hudson Valley. Through effective participation between the farmer and the consumer, farmers' markets will provide the foundation for empowered, balanced, and cohesive communities that rely on informed, equal, and educated transactions. Farmers' markets provide a relaxed social atmosphere to find all sorts of locally grown heirloom fruits and vegetables that have unique colors, textures, and tastes that cannot be found in factory-farmed industrial produce.

Figure 10.4 Interior views of the education facility

Source: Produced by author.

Perhaps most importantly, through the encouragement of **school group visits**, this facility will challenge the current separation of agriculture from education and recreation. Through the increasing recognition of the power of small actions, we can begin to take pride in our capacity to participate in a more humane and sustainable way of agriculture.

References

Broom, D., Sena, H., & Moynihan, K. (2009) Pigs learn what a mirror image represents and use it to obtain information. *Animal Behaviour*, 78.5: 1037–1041.

Emanuele, P. (2010) Antibiotic resistance. *American Association of Occupational Health Nurses*, 58.9: 363–365.

Evans, S. L., Graczyk, T. K., Graham, J. P., Price, L. B., & Silbergeld, E.K. (2009) Antibiotic resistant enterococci and staphylococci isolated from flies collected near confined poultry feeding operations. *Science of the Total Environment*, 407.8: 2701–2710.

FAO (Food and Agriculture Organization) (2007) *State of the World's Animal Genetic Resources*, FAO.

Gill, R., & Raine, N. (2014) Chronic Impairment of Bumblebee Natural Foraging Behaviour Induced by Sublethal Pesticide Exposure. *Functional Ecology*, 28.6: 1459–1471.

Hallmann, C., Foppen, R., Turnhout, C., Kroon, H., & Jongejans, E. (2014) Declines in Insectivorous Birds Are Associated with High Neonicotinoid Concentrations. *Nature*, 511.7509: 341–343.

Houpt, K. (1982) *Domestic Animal Behavior for Veterinarians and Animal Scientists.* Wiley-Blackwell.

Pimentel, D. (2003) World population, food, natural resources, and survival. *World Futures*, 59.145: 67.

Shulman, R. (2012) *Eat the City.* Crown Publishers.

Singer, P., & Mason, J. (2007) *The Ethics of What We Eat: Why Our Food Choices Matter.* Rodale Press, Inc.

Steinfeld, H., Gerber, P., Wassenaar., Castel, V., Rosales, M., & Haan, C. (2006) *Livestock's Long Shadow: Environmental Issues and Options.* Food and Agricultural Organization of the Unites Nations.

UNESCO (United Nations Educational, Scientific, and Cultural Organization) (2010) Energy Flow, Environment and Ethical Implications for Meat Production, UNESCO.

Young, R. (2003). *The Secret Life of Cows.* Farming Books & Videos Ltd.

11 Urban agriculture as a tool for sustainable urban transformation

Atatürk Forest Farm, Ankara

Kumru Arapgirlioğlu and Deniz Altay Baykan

Introduction: urbanization, food safety, urban health, and alienation

> Greening the 21st century city will improve our health, stabilize our economy and bring us all closer together as we meet in the garden.
>
> (Jack Smit, 2001)

There are several necessities for human life to continue: a place to shelter, food to feed, and an income to maintain all. To achieve better conditions and accomplish life many migrate to cities. The concentration of functions in cities while creating a productive and active environment for people, also generates many inequalities. Besides these inequalities, all societies have been trying to fight against two major problems in over-populated cities: poverty and environmental degradation.

Poverty and insufficient nutrition have become an important problem for many cities under the pressure of migration, unplanned growth, and urban sprawl. Urban sprawl threatens many environmental resources within the vicinity of cities, including valuable agricultural land, but also intensifies the problems of accessing safe, cheap, and nutritious food. Rapid urbanization and land speculation in urban territories causes land to be transferred into more overpriced uses such as housing, commerce, or even mega-public projects under the stroke of "urban renewal," and these invalidate agricultural and public land. Unplanned and dense urbanization of land, while creating threats to the basic foundation of human life and human health, causes alienation from the means of production, nature, and natural resources.

Cities rose and prospered as a result of fertile land and surplus food in history, but now they undervalue their main motive: nature. Major problems of urban life in relation to food supply in metropolitan cities appear to have changed: food has to be bought, is expensive, and usually unhealthy or not nutritious enough. Access to healthy and fresh food either becomes very limited or very expensive within the boundaries of cities. Though an important and necessary part of family survival and budget, food by itself becomes an expensive product and goes through a laborious process before it reaches the cities and city markets.

A very high percentage of population will be living in cities in the future; therefore arranging better food supply networks depends on how well the available urban land is re-allocated for agricultural production and supply of food. So reducing cost and enabling accessibility to cheaper and healthier food will be very important, especially for urban poor. Nevertheless keeping agricultural land productive will benefit society by accessing healthier food and organizing contact with land and production processes. Such transformations will also open new employment channels and opportunities through urban agriculture.

Urban sprawl also affects the land, which is valuable with its land cover and soil, and then its ecological and agricultural foundations. Thus without better management tools and better governance of urban land at the local scale and without getting adequate support from the national government and international organizations, all citizens but mostly the urban poor will continue to suffer more from environmental problems, undernourishment, and costly consumption as a result of unproductive use of public land. The academic arena must also provide research and case studies and develop guidelines and implementation principles to help out in the process.

Accordingly, many international meetings, documents, and conventions have been focusing on alleviating poverty, including, primarily insufficient nutrition and unhealthy living conditions in cities. Since the last decades of twentieth century, international organizations such as the Food and Agriculture Organization (FAO), Resource Centers on Urban Agriculture and Food Security (RUAF) Foundation, United Nations Development Programme (UNDP), United Nations Children's Fund (UNICEF), and the World Health Organization (WHO), with varied partnerships, have emphasized the importance of governance and management of urban land to curb these issues by building guidelines, leading case studies, and providing support to local and central governments, and to related parties. Scientific organizations, such as COST-Action UAE (Urban Agriculture Europe – European Cooperation in Science and Technology), draw attention to the importance of urban agriculture (UA) by innovative research, new methods (research by design), trans-domain networking, and through publications (for example the European Atlas on Urban Agriculture).

Many of the above organizations and researchers underline the importance of protecting, developing, and better management of urban agricultural lands, whether in the vicinity or in the city borders (De Wilt and Dobbelaar, 2005; McClintock, 2010; Mougeot, 2000, 2005, 2006; Viljoen, 2005; Viljoen and Bohn, 2014). Lastly, pressures from powerful stakeholders mostly override public land, public interest, and benefits for the urban poor; and commonly concerns over economic development and possibilities of higher returns usually weaken the governance of the benefits for the needy. As a result multi-actor, multi-functional supervision is highly stressed by the international community. The emphases on better governance and management provide a wider approach to issues, integrate many concerns, and have the power to assemble diverse groups.

Under the topic of Sustainable Urban Agriculture and Food Planning we concentrate mainly on governance of an urban "agricultural" land and here in this chapter we question and discuss the destiny of an urban farm, Atatürk Forest Farm (AOÇ),[1] as a case study in the heart of Ankara, Turkey.

After giving a brief comment on the historical development of the forest farm, the discussion will be built upon governance, productivity, and management of land in accordance with sustainable urban transformation, and will focus on new spaces and new social encounters through AOÇ's capacities. Along with this discussion, we also emphasize the empowerment of individuals as *active citizens* instead of being just onlookers. Our further aim is to question the case of AOÇ nationally and to create an international platform.

This chapter is an outcome of a research series carried out by the authors since 2012 and studio work by Arapgirlioğlu.[2] Believing that UA can be an important tool to revitalize developing cities and help to solve the above-mentioned problems, we have concentrated on AOÇ, which is under serious threat and holds potential as a case study. The case has involved the students of Bilkent University, Department of Urban Design and Landscape Architecture (LAUD) through fieldwork and assigned research. Several meetings and workshops on UA, organized by COST-UAE and AESOP,[3] also have supported this work and encouraged us to continue to work on UA and AOÇ.

A twentieth century utopia: the modern city of Ankara and Atatürk Forest Farm as a center for production

Mustafa Kemal Atatürk initiated AOÇ as part of a broader project in 1925 in Ankara.[4] The foundation of AOÇ was one of the important building blocks of the twentieth century utopian vision of Atatürk and the new Republic of Turkey, including shifting the capital from Istanbul to Ankara. At that time the major aim of building city farms was to provide the Anatolian cities with safe and healthy daily products, create production centers, and empower urban life with new public and leisure uses. Besides supplying food to the new capital and bringing new technologies and knowledge of agriculture to the new nation, it also aimed to produce a bond between young urban people and agricultural production, and guide "members of the Republic" to retain a "modern" urban life.

As a start, 2000ha of barren land by Ankara river was bought by M. K. Atatürk from the local owners (Öztoprak, 2006; Semiz, 2009) and as initiator of the idea; he focused and insisted on this barren land, which was specified as "unproductive" by most of the specialists invited to the capital; being restored, cultivated, and transferred into productive farm complexes containing places of application, education, and knowledge.

The initial model included the formation of a large city forest, creation of a natural park area, and initiation of new cultural services, which brought together the urban park and agricultural production in the capital. Being close to water resources[5] (Ankara river on the north border) and to abundant grain and wheat crops of Anatolia and to the railroad (on the south border) had been the major advantages of this barren land. Selecting a site close to the railroad made it easier for transportation of the raw materials from Anatolia and the distribution of products all over the country. With the establishment of an agricultural research center, the idea also contributed to the advancement of agricultural production in the region and in the rest of Anatolia including agro-technical research and new methodologies. Existing transportation axes (highway and railroad) supported the collection of the products and the demand-supply processes. Among other farms founded all over Anatolia, AOÇ was planned to be the model and leader (Semiz, 2009). A foreign technocrat, Herman Jansen, who had been nominated to prepare the master plan of the new capital, Ankara, also supported the idea of a city farm parallel to his garden city vision and applied the aspiration of Atatürk.

As a model, AOÇ had a multi-purpose use and aim. After 13 years of hard work to cultivate this land, M. K. Atatürk transferred his private ownership to the national treasury in 1937, the utopia became real, and the city started to reap the fruits of this institution on several levels. AOÇ then became a real model for farms all around Anatolia and also became a symbol of taming nature, bringing new insight to urban recreation (Kaçar, 2011; Semiz, 2009).

Since its foundation AOÇ has gone through three major development stages:

- **Foundation and formation stage**; acquisition of land, planning, managing and development of uses (1925–1937);
- **State farm stage**; enactment of AOÇ law and change of status to a state farm, to a major production area, and a public reserve (1938–1983);
- **Decline/shrinking stage**; AOÇ as an invaded land and as a subject of economic speculation and political return (1984–2014).

Foundation and formation stage (1925–1937)

During the foundation stage (1925–1937), the main layout of AOÇ as an urban design project was formed by Ernst Arnold Egli, who had also designed many buildings located on site (Alpagut, 2010). In 1936, Egli with Jansen, put together a modest model of modernity by sketching the new urban spaces and many other new buildings and functions within AOÇ[6] i.e. beer and malt factory, social housing for workers, a *hamam* (Turkish bath), a restaurant, a hotel with a swimming pool, a railway station, all offered many opportunities for citizens of Ankara. All these implementations with their social, economic, and recreational returns developed a new culture in Ankara (Kaçar, 2011).

Rehabilitation of the barren and swampland began as a fight against malaria and for the easement of further agricultural activities, and the greening of the area started with planting native trees and creating orchards. Production centers were founded to meet the needs of the city, such as greengrocery, dairy products, and ice. Livestock and apiculture programs started for stock breeding and bee fertilization; centers for producing agricultural crafts were fabricated, such as plows and iron tools. Altogether they established a new sector and a model for advanced agricultural production and know-how for the new Republic of Turkey.

State farm stage (1938–1983)

During the second stage as a state farm, AOÇ developed as a production area and as a public reserve. There were several planning acts and regulations put into practice in this period, and there were many improvements. Some of these improvements were: a milk and dairy products factory, built in 1957 with a contribution from UNICEF (the milk was supplied from different localities of Anatolia); and a wine factory was opened in 1961. The state farm was one of a kind in Anatolia, and it developed its capacity and brands during this stage including juice, pickles and vinegar production, honey and related products, especially during the period 1966–1967.

After the transfer of its ownership to the government treasury in 1937, AOÇ was under a lot of pressure to donate its land to other public institutions and usages (Öztoprak, 2006) and in response a law was implemented to protect its unity and status in 1950. However, despite such protective measures, AOÇ lost major quantities of its land to other public bodies through numerous changes made to AOÇ law (Figure 11.1).

Decline/shrinking stage (1984–2014)

The main characteristic of this third stage (1984–2014), other than sales and the renting of land, was the start of privatization in Turkey and the private sector's interest in dairy products and production. After 1990s with the enactment of privatization law, some of AOÇ lands owned by public institutions[7] were transferred to private enterprises; other functions of AOÇ also became the subject of privatization, which created new gaps in land and loosened authority and unity. As urban rents overpowered agricultural uses within the urban borders and as there were more risks in keeping land in agricultural production, AOÇ was considered as a reserve area and unfortunately was exposed to speculation, law suits, and was defeated by political greed and private enterprise.

Though AOÇ had been taken under protection as a first degree Natural and Historical Site by the Ankara Superior and Regional Council of Immovable and Cultural and Natural Property in 1992, by 2007 the planning authorization of AOÇ had been given to the Greater Municipality of Ankara, and by 2012 first degree natural and historical protection had been downgraded to third degree, opening the door to changes of land use and to construction of an eight lane highway dividing the land in two, a huge presidential house (450,000m^2),[8] a massive theme park[9] (one of the biggest in Europe), and expansion of the zoo area, causing extensive logging of trees (according to some sources adding up to 9,000 trees) and resulting in a break down of AOÇ's ecological integrity, land unity, and agricultural productivity.

During this stage, as the uses and the land had not been properly maintained by the responsible bodies, the citizens of Ankara started to lose their earlier bonds with AOÇ and to shift their interest to other newly developed recreation areas. This also led to the psychological break of ties with AOÇ and also with the main ideology

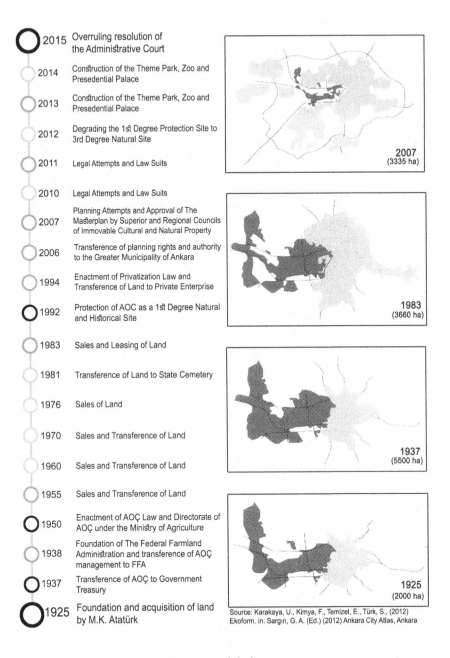

2015	Overruling resolution of the Administrative Court
2014	Construction of the Theme Park, Zoo and Presedential Palace
2013	Construction of the Theme Park, Zoo and Presedential Palace
2012	Degrading the 1st Degree Protection Site to 3rd Degree Natural Site
2011	Legal Attempts and Law Suits
2010	Legal Attempts and Law Suits
2007	Planning Attempts and Approval of The Masterplan by Superior and Regional Councils of Immovable Cultural and Natural Property
2006	Transference of planning rights and authority to the Greater Municipality of Ankara
1994	Enactment of Privatization Law and Transference of Land to Private Enterprise
1992	Protection of AOC as a 1st Degree Natural and Historical Site
1983	Sales and Leasing of Land
1981	Transference of Land to State Cemetery
1976	Sales of Land
1970	Sales and Transference of Land
1960	Sales and Transference of Land
1955	Sales and Transference of Land
1950	Enactment of AOÇ Law and Directorate of AOÇ under the Ministry of Agriculture
1938	Foundation of The Federal Farmland Administration and transference of AOÇ management to FFA
1937	Transference of AOÇ to Government Treasury
1925	Foundation and acquisition of land by M.K. Atatürk

2007 (3335 ha)

1983 (3660 ha)

1937 (5500 ha)

1925 (2000 ha)

Source: Karakaya, U., Kimya, F., Temizel, E., Türk, S., (2012) Ekoform. in: Sargın, G. A. (Ed.) (2012) Ankara City Atlas, Ankara

Figure 11.1 AOÇ: major stages of progress and decline

behind it. There was a major alienation towards what was happening tp AOÇ and to its future plans. Moreover, predominantly between 2005 and 2015, AOÇ faced serious land loss and experienced negligence of its agricultural land use and production (Figures 11.2 and 11.3).

Existing state of AOÇ: advantages, possibilities, and shortcomings

For more than 80 years AOÇ was a symbol of "reliable and healthy food," a center for urban leisure activities, and a widely used urban park. Since its foundation, with agricultural production and as a center of production, AOÇ has provided a wide range of dairy products (though with less market share now, it still does) not only to the Ankara metropolitan area, but also to its hinterland. AOÇ, up until the 1990s, with its advanced retailer network, know-how and production capacity, contributed to many agricultural establishments by purchasing and manufacturing products, providing financial and information support, and reaching many sales points. Now AOÇ network is catering to a wider range of consumers through its

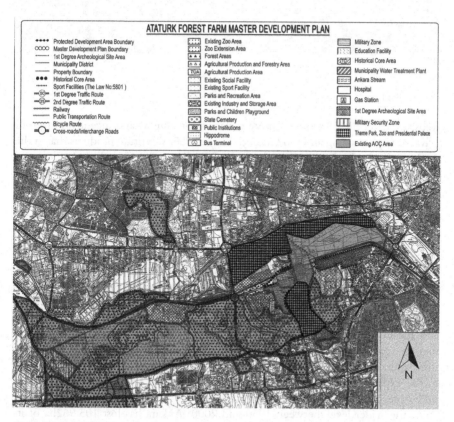

Figure 11.2 AOÇ master development plan

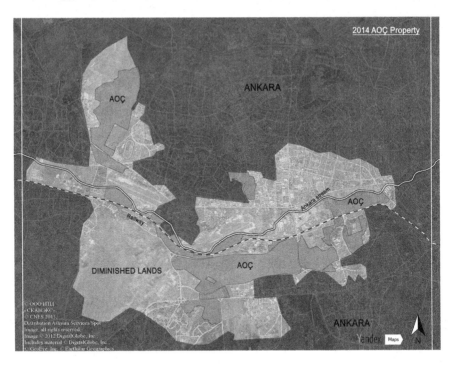

Figure 11.3 2014 AOÇ property

website.[10] After the 1990s, with more government support and incentives given to private manufacturing and food industry, some other substitute products shared the market. Along with lower investment in AOÇ and in agricultural production, and the supply of new diversified products on the market not only weakened AOÇ as a institution but also the goods of AOÇ lost their market share within the local and regional markets. This effect was also augmented by organizational and regulatory negligence towards AOÇ.

During the following years, the neoliberalization in urban sphere introduced a new mode of production to urban space and brought new experiences into the urban arena. Ankara as the second biggest city of Turkey, with governmental institutions and many other potential service sectors and functions, i.e. administration, education, health, and agriculture (until 2000s) (Table 11.1), became the target expansion area of the country after Istanbul. In this regard the approach of central and local authorities towards potential urban land drastically changed, with a massive construction boom within or in the vicinity of the city. Valleys, agricultural lands, wet lands, forest areas, treasury lands, and fringe villages were threatened by this expansion and new form of urban transformation. Eventually, AOÇ too was under pressure from this densification process.

At the wider level, between 1950 and 2000, Ankara has lost 105.962ha of its agricultural land with major changes in development plans (Sezgin and Varol,

Table 11.1 Economic sectors of Ankara (1970–2005)

Sectors	Years 1970 Active pop. (%)	1975 Active pop. (%)	1980 Active pop. (%)	1985 Active pop. (%)	1990 Active pop. (%)	2000 Active pop. (%)	2005 Active pop. (%)
Agriculture	263,989 **(36.0)**	313,419 **(32.3)**	271,664 **(27.6)**	288,097 **(26.4)**	200,191 **(18.1)**	223,488 **(16.2)**	94,000 **(7.3)**
Industry	79,589 **(10.8)**	136,686 **(14.1)**	130,490 **(13.2)**	141,330 **(13.0)**	147,758 **(14.3)**	184,897 **(13.4)**	303,000 **(23.5)**
Services	360,179 **(49.1)**	500,627 **(51.4)**	569,232 **(57.9)**	648,984 **(59.5)**	752,186 **(67.6)**	966,636 **(70.1)**	893,000 **(69.2)**
Other	29,882 **(4.1)**	20,935 **(2.2)**	12,881 **(1.3)**	11,919 **(1.1)**	3,190 **(0.01)**	3,678 **(0.27)**	–
Total	**733,639 (100)**	**971,667 (100)**	**984,207 (100)**	**1,090,330 (100)**	**1,103,325 (100)**	**1,378,699 (100)**	**1,290,000 (100)**

Note: Data include urban land.

Source: DIE, TÜIK-Turkish Statistical Institute (2005).

2012), and there has been a major shift from agriculture to service sectors in the economy (Table 11.1). Now with its population over 5 million, Ankara needs a new vision to sustain its future and feed its habitants.

Despite all its loss of land, devaluation of its identity and organizational unity, and societal and collective cultural status, AOÇ still has a rich and abundant potential to become a twenty-first century refuge in a densely urbanized city. There are 1000ha of agricultural land, green houses, and a research center, and there are over 1000 ha of meadow, commons and feed fields. Researchers confirm that almost 50 percent of AOÇ's agricultural land still has fertile soil (Dengiz and Keçeci, 2006). Within AOÇ boundaries, there is almost 20km of riverfront (Pekin, 2007) and 1000ha of forested area available to be used as a recreation area, including also many registered historical buildings and cultural sites (Table 11.2). Its existence as an open area is still important for Ankara.

The advantages and potentials of AOÇ include:

- An experienced and well operating establishment (since 1938);
- Its floating assets and product diversity (3300ha of land);
- Its large budget and profit from leasing and sales (an average annual budget of EU€23.5 million and a profit of €2 million from leasing and sales of products by 2011[11]);
- Good administration and skilled staff, with know-how and experience of at least 75 years of collection, production, and distribution of a variety of goods (around 25 brands);
- An advanced network of wholesale distributers and licensed retailers;
- Agricultural land, production, and established directorates of botanic production with a research center for saplings and good practice of Anatolian agricultural crops and saplings;

Table 11.2 Atatürk Forest Farm property and land use

Land use	Atatürk Forest Farm property (ha)							
	2004				2005			
	Irrigated land	Arid land	Total	Use (%)	Irrigated land	Arid land	Total	Use (%)
Cultivated land								
1. Arable fields	200.5	730.2	930.7	27.7	168.2	745.6	913.8	27.40
2. Horticulture and orchards	76.2	–	76.2	2.3	76.2	–	76.2	2.28
Total (1+2)	**276.7**	**730.2**	**1006.9**	**30.0**	**244.4**	**745.6**	**990.0**	**29.68**
3. Meadow, commons and feed fields	14.4	20.0	34.4	1.0	14.4	20.0	34.4	1.03
Total (1+2+3)	**291.1**	**750.2**	**1041.3**	**31.0**	**258.8**	**765.6**	**1024.4**	**30.72**
4. Other lands:								
Occupied land	–	98.3	98.3	2.9	–	75.0	75.0	2.25
Forest and parks	–	1150.4	1150.4	34.4	–	1150.4	1150.4	34.49
The zoo	32.0	–	32.0	1.0	32.0	–	32.0	0.96
Rental property	–	688.1	688.1	20.6	–	703.2	703.2	21.08
Vacant and unused lots, roads, and canals	–	338.6	338.6	10.1	–	350.1	350.1	10.50
Total (4)	**32.0**	**2275.4**	**2307.4**	**69.0**	**32.0**	**2278.7**	**2310.7**	**69.28**
Overall total	**323.1**	**3025.6**	**3348.7**	**100.0**	**290.8**	**3044.3**	**3335.1**	**100.0**

Source: Greater Municipality of Ankara (2006).

- Location and accessibility; it is in the very center of the city allowing easy access of people and goods;
- Water resources; Ankara river, besides its recreational potential with its long riverfront (20km riverfront), holds many other potentials including becoming a cooling corridor for Ankara, strengthening its ecological integrity;
- A vast forested area holding various species, with 2 million trees planted after its foundation;
- Building stock, infrastructure, and uses, with examples of the very early Republican architecture, holding an important stock of cultural heritage and collective memory.

The disadvantages and shortcomings of AOÇ include:

- AOÇ has been neglected for more than two decades and very few financial and personnel resources are allocated from its budget to agricultural production, to management, and to the maintenance of its assets;
- Many of its uses have been either diminished (the zoo), shut down (wine factory), or ineffectively used;

- Ankara river and its waterfront have been neglected, with no water treatment and used as a dumping ground;[12]
- The recreational and agricultural activities have lost their previous significance, efficiency, and liveliness, which also caused many businesses to close down in the vicinity of the site;
- What is left of AOÇ is under the threat of ongoing exploitation.

However, AOÇ, with its numerous potentials and with a new organizational perspective could overcome its disadvantages, and has great capacity to become an efficient "urban agro-open space system," a place for improving urban agriculture, a new center for food planning and management, and to become an urban open space for citizens and for reconstructing the twenty-first century Ankara as a productive and self-sufficient city.

In this respect, governance, planning, participation (involvement of actors), and decision-making processes are essential for generating a new perspective for AOÇ, and considering urban agriculture as a new tool for urban transformation is essential for sustaining its core values. All concerns related to mismanagement of AOÇ could be overcomed by improved governance and social/cultural gains reaped under a new act, an organization, and a concept, combined together under the concept of an "urban agro-open space system" that integrates urban agriculture with the concept of an agro-park.

A joint model "urban agro–open space system": integration of urban agriculture with the concept of agro-park

Regional or global initiatives hold several perspectives for sustaining cities. Their concerns are twofold: one is the ineffective use and management of resources. Second is the high concentration of population in cities. It is predicted that there will be 1.5 billion people living in the periphery of cities by 2020, mostly unemployed and lacking access to nutritious food. One of the key instruments to overcome this troubled future is to create self-sufficient cities that can feed themselves. Promoting urban agriculture and building aagro-parks within cities arose as core solutions in many cases.

Urban agriculture (UA), since the early 2000s, has been the focus of governments, organizations, and scientific societies as a solution to many of the problems mentioned above. It has been cited by many (McClintock, 2010; Mougeot, 2000, 2005, 2006; Viljoen, 2005; Viljoen and Bohn, 2014) that UA improves food security, creates jobs, balances ecology of the city, and has many more advantages.

UNDP (1996, 2001) defines UA as a foundation system, a supportive act, and an organization for urban economic activities, urban recreation, and urban aesthetics. According to other sources, UA has to be differentiated from informal urban agriculture, which exists mostly in urban ghettos or *gecekondu* areas,[13] and from privately owned community gardens. Principally UA is considered as an organized form of agricultural network for cities and their vicinities. It is mostly carried out on public land, and allocation of production is operated by cooperatives or public organizations, thus providing profit not only to society in general but a greater profit to the producers and consumers who are involved in this activity, and it also helps develop confidence.

According to international documents (UNDP, 1996, 2001; FAO, 1996, 2006, 2011; van Veenhuizen, 2006) UA has to be:

- Defined within the frame of organizational model;
- Taking place within social development policies;
- Developed to create income and decrease unemployment;
- Aimed at food security, developing necessary precautions and controls;
- Have defined spatial boundaries;
- Involve environmentally sensitive waste and resource management;
- Involve new technology, considering organic agriculture;
- Contribute to the urban open space networks, landscape aesthetics, and urban recreation.

In the case of AOÇ, many of the above mentioned criteria and characteristics fit with the definitions that empower the idea of UA.

According to Mougeot (2000), different from other types of production, UA activates multiple actors, includes local and central government, and creates interactive communication. These peculiarities once more validate UA as a regenerative tool to reform a twenty-first century city, as illustrated by the example of AOÇ and Ankara.

UA, through these practices, could become a helpful tool also to control three important realities of dense urbanization and neoliberal urban politics: the use of agricultural land for non-agricultural purposes; environmental problems and degradation due to high population and urban sprawl; and the isolation of individuals and society from land and nature.

Figure 11.4 AOÇ farmlands, forestry and orchards

Recently, with growing interest, governments and non-governmental organizations (NGOs) have started to invest in UA, not only to lower food costs but also to reclaim urban lands for the use of the society.

Agro-parks are another approach to managing such lands, establishing a multipurpose urban use and a program specially focusing on agricultural activities and marketing. They not only work as agricultural production spaces but also for the production of socially active and attractive urban spaces, which have many advantages. They are self-contained systems, reduce transportation costs, bridge gaps between different social groups, and generate social, economic, and ecological benefits (De Wilt and Dobbelaar, 2005). Thus, the agro-park concept coupled with UA may advance diverse motives into regenerating urban agricultural land and uses. Nevertheless in rural agricultural systems, it has also been proved that a combination of uses and multiple functions on agricultural land multiplies benefits. For example agriculture, forestry, and livestock uses together bring more efficiency of use and productivity by lowering distance between uses and users, by creating closer communication networks, and by easing interaction of different sub-groups of the larger sector (Frey *et al.*, 2012).

In Ankara, there are many actors and coalitions acting against the exploitation of public lands by urban transformation projects. In the AOÇ case there are two major opposing parties: one is the members of The Union of Chambers of Turkish Engineers and Architects (UCTEA), and second is the NGOs. In Turkey members of UCTEA[14] have always been powerful actors and defenders of public lands and ecological values. However after the 2000s some environment-focused NGOs[15] also started to fight for major environmental issues. These actors have been fighting back all major changes in the case of AOÇ through protests, law suits, and by organizing meetings and conferences aiming to create public involvement and awareness, and also to create public pressure on the politics of growth. A recent resolution of the Administrative Court, which is a result of the efforts of these actors, has overruled all changes made recently on AOÇ plans by the Greater Municipality of Ankara.[16]

Loss of cultural identity is also another important debate: the inhabitants of Ankara for more then half a century felt themselves to be an important part of AOÇ by consuming its products and spending their leisure time and weekends on grounds of AOÇ. A survey done shows that AOÇ still has an important share in citizens' memories and worries (Açıksöz and Memlük, 2004). Therefore efforts to regenerate AOÇ will provide a chance to create a new social setting and create a socially productive bond among citizens of Ankara. A new beginning for AOÇ will generate a synergy echoing that of a century ago and moreover, redeveloping AOÇ with multiple uses will give a chance to regain the memories and knowledge of AOÇ for the generations who have lost their contact with it. Along with these efforts, reviving AOÇ's agricultural identity by promoting UA, renewing its waterfront for recreational purposes, and giving it new contemporary uses as an agro-park, will protect the land from exploitation of other forces and actors, and will give it a chance to survive and enhance its ecological and cultural identity.

In this respect, planning and design are other necessary steps to be taken towards

Figure 11.5 The Presidential Palace (centre) and Marmara Kiosk (right) designed by E.A. Egli in 1928 for Atatürk as a farmhouse (seen from the highway crossing AOÇ)

achieving these aims and objectives. Planning and design are important mediators between political and mental activities, turning thoughts into useable spaces (Lefebvre, 2007). Through practice, experience, and social activity, new forms of public spaces and uses will be accomplished by these acts and AOÇ could regain its social character and agricultural activities, together with enhancing its recreational and ecological potential.

Though it may seem like wishful thinking under present conditions, it is unavoidable that local government and central government and AOÇ administration must join forces with concerned NGOs, unions, chambers, and universities to create a more productive approach, effective interaction, and participation over the issues. This will bring many benefits, including public productivity, and empowering future generations by designing a self-sufficient city.

Conclusion: a model for a sustainable future and a space of social production

New policies and programs can help to improve the conditions of publicly owned lands. UA and the agro-park concept can create new opportunities, leading to the development of a more productive, actively, and freely experienced urban public open space. By doing so, multiple benefits could be gained in the case of AOÇ, including:

- Restoring the productivity of degraded agricultural land;
- Improving food productivity and effective use of land;
- Overcoming the disruption of the riverfront ecology;
- Generating new economic relations and employment networks;
- Strengthening the local economy and enhancing a sense of community and networks;
- Offering space for sharing and producing together;
- Bridging the gap between producer/consumer, citizen/land, youth/nature;
- Creating a platform for new innovations and advancements;
- Centralizing many uses and productions;
- Combining and clustering different sectors at one center;

- Reducing transportation costs and lowering fuel costs;
- Creating work as a process of social and individual development;
- Creating new meanings, values, and ways of life;
- Enhancing human physical and psychological health;
- Highlighting urban landscape with greening and maintaining open spaces;
- Working as an aesthetic stimulus for the citizens;
- Integrating agricultural uses with urban activities.

Future work to be done in AOÇ could involve two phases: promoting UA as an act and network for Ankara, together with advocating the agro-park concept, which will progress AOÇ's industrial, recreational, and cultural characteristics.

UA, integrated with the concept of agro-park, is proposed as a strategy for the renewal of AOÇ and as an urban agro-open space system and could follow these steps:

- Collecting and assessing relevant data on Ankara's socio-economic profile and unemployment data;
- Collecting relevant data, building up an inventory of existing capacities of AOÇ and assessing its potential (natural, social, economic) for future developments;
- Making sure that UA takes place within national and international communication networks in relation to food security and job creation;
- Providing a planning assessment of UA;
- Drawing attention to UA within planning practices and social policies;
- Assessing Ankara's agricultural sector and related industrial production from the perspective of UA;
- Initiating special incentives for UA within the borders of Ankara and for the structure of the budget within AOÇ;
- Assessing the urban hinterland, and collecting data on additional potential sites appropriate for UA in the close vicinity of Ankara;
- Initiating a new financial, administrative model backed up with integrative planning and design scales;
- Establishing an UA research and development center to work in coordination with the existing research center;
- Improving the existing AOÇ workforce, incorporating UA as practice;
- Involving organic agriculture awareness and practices;
- Involving many different interest groups/actors to become a part of the practice of UA;
- Exploring possibilities of pooling various resources by involving various groups such as the young, female, and disabled;
- Considering additional and innovative uses to attract more partners;
- Preparing weekly, monthly, and yearly activity programs/charts, and reducing conflicts by promoting interaction between different uses and users;
- Integrating all means of media for publicity of UA and the activities of AOÇ to create higher interest and involvement;

- Introducing many innovative and environmental practices within AOÇ, integrating with UA;
- While preserving historical and cultural sites, developing new activity centers for all users with UA;
- Introducing an open and semi-closed bazaar area for the interactive exchange of goods, providing cheaper, healthier, easily accessible food;
- Providing cheap, easy, and safe access to the UA market area;
- Taking notice of feedback and monitoring of the process;
- Structuring a center for promoting productivity, for awareness of food security, production processes, and environmental issues at every level.

All of the above components could not be realized without: proper governance of existing data and uses by planning and design teams; governance of resources and finance by the administrative organization supported by an expert team; governance of activities and people by public relations and social worker teams; governance of productivity by scientific teams. All of which could only be accomplished only by an higher organization, a multi-layered, interactive, and collaborative (such as an advisory board). This will also enhance and support the know-how of local and central government.

How to structure and to oversee is essential, in order to succeed in a multi-purpose, multi-layered arrangement, which needs interactive communication promoting cooperation; collaboration and combining forces, by uniting know-how of and involving all related actors (ministries, chambers, universities, NGOs) in the case of AOÇ.

Though AOÇ's existing organization with its experience of over half a century will create an important guide and supervisory role, we propose a new governance model for the revival and better management of AOÇ. In this model UA, being the umbrella concept, will trigger new progress and a new sprit for the land of AOÇ. For better functioning and efficiency, we propose a multi-layered, multi-purpose use that is the agro-park. Together both UA and the agro-park will form an urban agro-open space system, which in turn will strengthen Ankara's open space system. Therefore joining two powerful ideas for the revitalization of urban agricultural land will help bring about higher productivity in AOÇ.

This new organization has to have the freedom to act, creating its own guidelines and rules, and must be managed professionally to ensure satisfactory performance, while a committee of experts (members from the local and central governments, universities, chambers, and NGOs) will advise and oversee the model's processing and productivity. This model has to involve an independent non-profit organization that will be able to sustain a self-sufficient economic and administrative system in order to be free from political and rental pressures on one hand, and on the other an adequate environmental recycling system and environmentally sensitive approach to lower the pressures of rapid and dense urbanization.

Existing AOÇ law and related regulations must be restructured according to the new needs and to meet the new management design. Using its capacities and improving many existing functions of AOÇ will multiply its productivity and open space quality for new challenges. AOÇ management and law should be founded on three basic platforms: restoring the existing agricultural land; the maintenance of the neglected waterfront and adjacent areas; and the rehabilitation of its historical and cultural sites. These three developments should also provide additional functions where inhabitants of the city will meet and the productivity of social space will be challenged.

Proposed activities and uses will primarily consist of farming, recreation, and marketing. AOÇ will continue to function as a multi-purpose farm, while also offering many environmentally friendly leisure and commercial activities that will lead to exchange of goods and ideas. By enhancing its green character, keeping its existing land and many of its functions will not only help to sustain its own ecological balance but will also support urban ecological balances and nurture ecological corridors of the city. Despite all the negative developments on AOÇ land, we are still hopeful that the politicians will hear many of our calls for environmental awareness, food security, and healthy and productive life-styles, and anticipate that they will cooperate to overcome many obstacles resulting from over population and rapid urbanization. We believe that publicly owned lands are important assets and they ensure our ability to sustain a healthier life in cities. We hope that the proposals and discussions on public land, such as in AOÇ case, will offer new challenges and perspectives in the creation of more productive social spaces and self-sufficient cities.

All these features, the land, the people, and the functions, if governed well and wisely, will enhance not only the production capacity of the land but also provide citizens with a wide range of open spaces that will allow them to be more productive physically and psychologically. If a project such as AOÇ is governed consistently, as happened at the beginning of the twentieth century, it will always find its direction and receive approval from the majority of people and will also be rewarding for politicians in terms of future hopes and self-assurance. The AOÇ will therefore become an important refuge for many giving them a chance to regenerate themselves, preparing for the future, helping to produce new interactions, and new visions to maintain a perspective of environmental awareness and to reunion with nature.

Notes

1   Atatürk Orman Çiftliği in Turkish and referred to as AOÇ in the text.
2   The authors of this chapter, D. Baykan, K. Arapgirlioglu, with B. Batuman and G. Çulcuoglu, presented the paper *Urban Agriculture and AOÇ as an Urban Transformation Strategy* at the Symposium Atatürk Forestry Farm and the Future of Ankara, organized by TMMOB Chamber of Architects, ANKARA, 8 October 2012; and AOÇ has been studied as a case study in LAUD 481 Landscape Ecology course, taught by Arapgirlioğlu in 2012–2013, involving students in the process.
3   6th International AESOP Sustainable Food Planning Conference.
4   The founder and the leader of the new Republic of Turkey and also the initiator of similar farms all around Anatolia as a part of the Republic's foundation.
5   Including also the rivers of Cubuk, Macun, Incesu, Bend, and Kutugun (Semiz, 2009).
6   http://aocarastirmalari.arch.metu.edu.tr/; www.goethe.de/ins/tr/ank/

7   The lands owned by Sümerbank (former State Economic Enterprise on textiles), Directorate of Tekel/Monopoly (former State Economic Enterprise on Liquor and Tobacco Products) and Agricultural Equipment Institute were all sold to the private sector and are private islands within AOÇ (www.zmo.org.tr/genel/).
8   www.hurriyet.com.tr/ekonomi/27688912.asp.
9   Total 1,200,000 sq. m. including 100,000 sq. m. of indoor space (www.melihgokcek.com/proje-detay/anka-park-25.html)
10 www.aocsatismagazasi.com/ with the slogan of "natural, healthy and reliable."
11 Estimated to be around €37 million in 2015 budget, profits of €3.3 million by 2013 (www.meclishaber.gov.tr/).
12 Recently under construction due to new land-use changes and projects.
13 *Gecekondu*, Turkish origin, are areas of squatter houses, meaning "houses done overnight."
14 The Union of Chambers of Turkish Engineers and Architects (UCTEA), Chamber of Landscape Architects, Chamber of Architects, Chamber of Planners, Chamber of Environmental Engineers, Chamber of Agricultural Engineers (with recent changes in regulations, their budgets have been cut down as a result of their opposing attitudes).
15 Association of Ecology Collective of Contemporary City of Ankara.
16 This recent decision of the Administrative Court made all the changes and construction to date questionable, including the construction of the presidential palace, theme park, and expansion of the zoo, which were all previously protected against such actions under AOÇ law and other related regulations.

References

Açıksöz, S., Memlük, Y. (2004) Revaluation of "Atatürk Orman Çiftliği" with Respect to Urban Agriculture, *Tarım Bilimleri Dergisi* 10(1): 76–84.
Alpagut, L. (2010) Traces of Ernst Egli in Atatürk Forest Farm: Site Planning, Beer Factory, Housing and the "Traditional" Hamam, *METU JFA* 27(2): 239–264.
Chamber of Architects Ankara Branch TMMOB (2012) *Ankara City Atlas*, Ankara.
Baykan, D., Arapgirlioğlu, K., Batuman, B., Çulcuoğlu, G. (2013) AOÇ and Urban Agriculture as a New Urban Renewal Strategy. The International Symposium on Atatürk Forest Farm and The Future of Ankara. Unpublished paper.
Dengiz, O., Keçeci M. (2006) Land Assessment for Soils of the Atatürk Orman Çiftliği Based on Their Agricultural Uses, *J. of Fac. of Agric. OMU* 21(1): 55–64.
De Wilt, J., Dobbelaar, T. (2005) Agro parks: the concepts, the responses, the practice. Innovation network. In: Van der Schoor L. (ed.) (2005) *Report number 05.2.095E*, Utrecht: Drukkerij Rosbeek BV.
FAO (1996) *Urban Agriculture: An Oxymoron? The state of food and agriculture*, FAO, Rome.
FAO (2006) *Urban Agriculture: For Sustainable Food Alleviation and Food Security*, FAO, Rome.
FAO (2011) *The Place of Urban and Peri-Urban Agriculture (UPA) in National Food Security Programs*, FAO, Rome.
Frey, G. E., Fassola, H. E., Pachas, A. N., Colcombet, L., Lacorte, S. M., Renkow, M., Pérez, O., Cubbage, F. W. (2012) A Within-Farm Efficiency Comparison of Silvopasture Systems with Conventional Pasture and Forestry in Northeast Argentina, *Land Economics* 88(4): 639–657.
Greater Municipality of Ankara (2006) *Report of the Protection Plan for AOÇ*, Ankara.
Kaçar, D. (2011) A unique spatial practice for transforming the social and cultural patterns: Atatürk Forest Farm, *METU JFA* 28 (1): 165–178.
Lefebvre, H. (2007) *The Production of Space*, Translated by Donald Nicholson-Smith. Blackwell.

McClintock, N. (2010) Why farm the city? Theorizing urban agriculture through a lens of metabolic rift, *Cambridge Journal of Regions, Economy and Society* 3: 191–207.

Mougeot, L. J. A. (2000) *Urban Agriculture: Definition, Presence, Potentials And Risks, Growing Cities, Growing Food: Urban Agriculture on the Policy Agenda: A Reader on Urban Agriculture*, DSE CTA Sida GTZ, Havana: Cuba.

Mougeot, L. J. A. (2005) *Agro polis: the Social, Political and Environmental Dimensions of Urban Agriculture*, Canada: IRDC.

Mougeot, L. J. A. (2006) *Growing Better Cities: Urban Agriculture for Sustainable Development*, Canada: IRDC.

Öztoprak, I. (2006) *History of Atatürk Forest Farm*, Atatürk Araştırma Merkezi, Üç S Basım: Ankara.

Pekin, U. (2007) Development of Urban River Corridors and Concept Greenway Plan of Ankara Stream, Unpublished Ph.D. Thesis, Ankara University Graduate School of Natural and Applied Sciences Department of Landscape Architecture.

Semiz, Y. (2009) Ataturk's Farms and Their Transfer to the Treasury, Selçuk University, *Türkiyat Araştırmaları Dergisi*, 166: 155–192.

Sezgin, D., Varol, Ç. (2012) The Effects of Urban Growth and Sprawl on the Misuse of Fertile Agricultural Lands in Ankara, *METU JFA* 29(1): 273–288.

UNDP (1996) *Urban Agriculture: Food, Jobs and Sustainable Cities Habitat II Series*, 1st Edn, UNDP, New York.

UNDP (2001) *Urban Agriculture: Food, Jobs and Sustainable Cities, Habitat II Series*, 2nd Edn, UNDP, New York.

Van Veenhuizen, R. (ed.) (2006) Cities Farming for the Future: Urban Agriculture for Green and Productive Cities, RUAF Foundation, IDRC, IIRR and ETC Urban Agriculture: Philippines.

Viljoen, A. (ed.) (2005) *CPUL's Continuous Productive Urban Landscapes: Designing urban agriculture for Sustainable cities*, Elsevier,

Viljoen, A., Bohn, K. (eds) (2014) *Second Nature Urban Agriculture: Designing Productive Cities*, Routledge, Taylor & Francis: London and New York.

12 Elevating urban agriculture and city resilience

Making the case in the City of Los Angeles

Laura Sasso

Introduction: linkages between climate change, planned adaptation and resilience

Climate change impacts include shifts in the timing of seasons, extremes in regional temperatures, intensification of storm events and modification of precipitation patterns. By 2050 it is estimated that 67 percent of the total global population will live in urban areas (United Nations 2012). Due to the concentration of people, cultural assets and infrastructure, cities will be the stages where high-risk climate change impacts unfold.

In 2007 the Intergovernmental Panel on Climate Change (IPCC) defined planned adaptation as "adaptation that is the result of a deliberate policy decision, based on an awareness that conditions have changed or are about to change and that action is required to return to, maintain, or achieve a desired state" (Perry, *et al*. 2007). Planned adaptation cannot eliminate all climate change risks, but it can support resilience. The Stockholm Resilience Center defines resilience as "the capacity of a system, be it an individual, a forest, a city or an economy, to deal with change and continue to develop" (Moberg and Simonsen 2014). Adapting cities with resilient strategies bolsters the ability of its inhabitants to manage and live with environmental change. Therefore, effective planned adaptation in cities is an opportunity to improve the performance of urban environments and to mitigate existing social inequities in the near and long term.

Multi-criteria evaluation: contextual research for the performance indicators

Energy use in buildings

A key contributor to climate change is greenhouse gases (GHGs), a byproduct of energy consumption. In 2010 the U.S. building sector consumed 41.1 percent of the nation's total energy, which was higher than the transportation sector with 28.1 percent (U.S. DOE 2010). Furthermore, it is projected that total GHG emissions from buildings will grow faster than any other sector over the next couple of decades in the U.S. (Larsen, *et al.* 2011). A 2007 report from the IPCC concluded that "over the whole building stock the largest portion of carbon savings by 2030 is in retrofitting existing buildings" (Levine, *et al.* 2007). The IPPC report opens the door for extensive research and comprehensive implementation of strategies designed to reduce GHG emissions from existing building stocks.

While research on efficient strategies for new construction addresses one component of the urban fabric, it neglects to address the potential sizable energy saving from the exiting building stock. Significant vulnerabilities in the existing building stocks of cities will persist in the near-term if they are not integrated into adaptation planning processes. For example, in the United Kingdom the rate at which the older building stock is being replaced with more efficient new buildings lags behind the rate of change that is needed to reduce GHGs to targeted levels (Castleton, *et al.* 2010).

Implementing living systems on roofs can decrease energy use in buildings. In the United States, overall energy consumption in buildings is predominantly the result of heating and secondarily of cooling (U.S. DOE 2010). Retrofitting existing buildings with green roofs affects energy consumption in three critical ways: net energy use, peak energy loads and operational costs (Sailor, *et al.* 2011). Building type, vegetation height and the substrate depth of green roofs influence the potential for energy savings from a retrofit (Sailor, *et al.* 2011) (Simmons, *et al.* 2008). In a building, the upper floor, directly below a green roof, has the greatest potential for a reduction in cooling load (Spala, *et al.* 2008).

The City of Los Angeles is a pertinent example of Post-World War II urban sprawl. Low-level buildings dispersed over a wide area dominate the urban fabric. In Mediterranean climates, such as Los Angeles, green roof retrofits achieve the greatest change in energy use through reducing cooling loads during the summer season. Therefore, the combination of the preexisting network of low-level buildings and the high cooling loads in Los Angeles make a strong case for the use of green roofs to reduce energy use in existing building stock.

Urban microclimates and rooftop surface temperature

Climate change is exasperating heat-related vulnerabilities in many regions (IPCC 2012). The removal of heat-mitigating vegetation and the introduction of low albedo surfaces have elevated heat-related concerns in cities (Whitford, *et al.* 2001). An indirect result from greening building envelopes is the cooling of air temperatures through evapotranspirative processes. For green roof and vertical greening systems, Alexandri and Jones (2006) found that the thermal comfort results were highest in hotter, drier climates and that rooftops have a greater impact on the overall urban climate due to high solar radiation exposure (Alexandri and Jones 2006).

The city of Los Angeles is situated in a Mediterranean climate. Between 1981 and 2010 the average mean max temperature for August was 29.1°C (84.4°F) and for February 20.3°C (68.6°F) (NWS Forecast Office 2015). The latest research predicts that the region will experience an increase in temperature of around

1.95°C (3.5°F) by 2050 (Union of Concerned Scientists 2012). Heat-related health issues, e.g. heat stroke and exasperation of existing chronic diseases, are anticipated to be one of the city's greatest climate change vulnerabilities. Due to the year-round warm, dry climate, adapting the existing building stock in Los Angeles with productive landscapes has a great potential to reduce rooftop surface temperature, which would help to mitigate extreme-heat issues.

Stormwater management

Climate change is bringing more extreme storm events where greater amounts of precipitation are falling within a shorter timeframe (IPCC 2012). Impervious surfaces, a common characteristic in urban areas, elevate stormwater-related risks. Retrofitting roofs with living systems reduces the amount of impervious surfaces in developed areas. Gill *et al.* (2007) modeled the potential impact of ubiquitously greening roofs in three town centers in Greater Manchester in England. The large-scale simulated research resulted in about a 13 per cent mean reduction of stormwater runoff for a 28 mm (1.1") storm event (Gill, *et al.* 2007). Simmons and colleagues (2008) call for more research on the performance of different green roof typologies with different material compositions and substrate depths (Simmons, *et al.* 2008).

In the Los Angeles area total precipitation between 1981 and 2010 averaged 37.92 cm (14.93") (NWS Forecast Office 2015). Precipitation climate change models for the Southern California region are still uncertain as to the impact the phenomenon will have on the distribution of precipitation in the city. It is possible that there will be a decline in annual precipitation or it could be that there will be an intensification of storm events. A network of productive green roofs would reduce urban runoff, decrease flooding risks and support the production of food.

Urban access to healthy food

In 2012, the IPCC published a special risk report, which highlighted that the nexus of expanding populations and climate change is raising the issue of food security to a major global concern (IPCC 2012). The projected impact of climate change on the production of food is predicted to be unfavorable. First, increases in pests and diseases are anticipated to trump any potential increases in production yields as a result of higher CO_2 levels (Long, *et al.* 2006). Second, it is anticipated that elevated CO_2 concentrations will result in a decline in food quality (Bloom, *et al.* 2014). Irregularities in climate patterns have already caused regional food insecurity issues that have had global repercussion. A pertinent example occurred in 2010 when Russia's post-drought ban on grain exports resulted in food insecurity issues in Egypt (Hassan 2010). Custot and colleagues conclude that there is an urgent need to adapt urban food systems to bolster urban resilience (Custot, *et al.* 2012).

Urban dwellers living in food deserts are more vulnerable to fluctuations in food supply chains and they have a significantly greater risk of developing food-related health risks. Urban food deserts are neighborhoods that lack nutritious food options. The prevalence of fast food restaurants and convenience stores with empty calorie food options results in higher rates of obesity and food-related risks. In the County of Los Angeles approximately 800,000 residents live in one of the largest urban food deserts in the U.S. Developing a comprehensive urban food system that includes the production of and access to healthy food grown in the neighborhoods is an opportunity to support social-ecological resilience on a daily basis in Los Angeles.

In a food desert the probability of developing metabolic syndrome, a diagnosis of at least three of the five shared risk factors for heart disease and diabetes, is elevated. Metabolic syndrome correlates with twice the risk of developing heart disease and quintuples the probability of being diagnosed with diabetes (NIH 2011). Findings from a 2013 report from the Los Angeles County Department of Health concluded that in South and East Los Angeles an average of only 12 percent of adults consumed the federally recommended five or more servings of fruits and vegetables per day. These were the lowest consumption levels in the county. In the same year the percentages of obese adults in South Los Angeles (33 percent) and East Los Angeles (30 per cent) were over three times higher than in the more affluent community of West Los Angeles (LA County DPH 2013).

These diet-related health indicators illustrate inequities in nutritious food access in Los Angeles and its detrimental health implications. Notably, it is possible to prevent or postpone metabolic syndrome with adequate physical activity and a healthy diet.

Rooftop strategies

This research is focused on assessing three rooftop strategies: black synthetic rubber membrane (BM), green roof agriculture (GRA) and hydroponic greenhouse rooftop agriculture (HGRA). BM technology has been a common rooftop strategy for low-slope buildings for more than 40 years (EPDM Roofing Association 2014). Therefore, it is used as the baseline strategy to compare and contrast the two retrofit strategies in the study.

One of the most renowned examples of GRA is the Brooklyn Grange. It is a 1 hectare (2.5 acre) farm, split over two rooftops, and it produces over 22,680 kilograms (50,000 pounds) of organically grown food annually (Brooklyn Grange 2015). In addition to the production of food, the Brooklyn Grange partners with a local non-profit to supports the development of job skills through the Refugee and Immigrant Program. ØsterGro in Copenhagen is a 600 m² (1968.5 ft²) farm that produces vegetables, honey and eggs for local families as well as supports community engagement opportunities (ØsterGro 2015).

A leading example of HGRA is Gotham Greens in New York City and another large-scale rooftop project is under development in Chicago. The 18,288 m² (60,000 ft²) greenhouse in Brooklyn grows over 100 tons of green leaf vegetables (Gotham Greens 2015). In Canada, Lufa Farm's Laval site, with 13,106 m² (43,000 ft²) of greenhouse growing area produces 120 tons of produce each year (Lufa Farms 2015).

Both forms of rooftop farming produce food, but each system has distinct ecological and social benefits. HGRA can produce a greater quantity of certain types of vegetables when compared to the GRA. However viewing a retrofit strategy through a single lens – yield – provides a myopic view for decision-making processes. The near term and long term vulnerabilities and goals of the community are essential as well. For example, the production methods utilized in GRA have a lower barrier to entry making this strategy better suited for shared public-private spaces that fosters community development. Each strategy has advantages and disadvantages and it is essential to understand each to support informed, effective planned adaptation processes.

Objectives

Researchers from a range of disciplines call for studies focused on the potential impacts from large-scale implementation of green roofs in a range of climates (Carter and Butler 2008; Oberndorfer, *et al.* 2007). Furthermore, scientists advocate for research that utilizes multi-criteria assessments that include social, ecological and economic parameters to build comprehensive understanding about the potential of greening building envelope strategies (Carter and Butler 2008; Levine, *et al.* 2007; Pearson, *et al.* 2011). Therefore, this research builds on a robust set of interdisciplinary research on greening building envelopes and aims to develop understanding about the large-scale potential of rooftop farming.

The primary research objective is to estimate the large-scale potential impacts of retrofitting existing rooftops with two productive landscapes strategies through the evaluation of four performance indicators: building energy use, surface temperature, stormwater runoff and yield.

Furthermore, this research strives to quantify and connect potential crosscutting benefits from large-scale rooftop farming in order to spur the development of supportive policy frameworks. It is designed with a replicable research methodology, which could be used to facilitate food-based adaptation planning in other cities. Food-based planned adaptation strategies provide opportunities to bolster environmental performance, to mitigate existing inequities in food access and to support human health.

Research parameters: three modeling parameters for the case study

Model unit: flat roofed commercial buildings

Only flat roof commercial buildings in the existing building stock are assessed. In the U.S. the most comprehensive, publicly available set of data on commercial buildings is the Commercial Buildings Energy Consumption Survey (CBECS). Existing research from CBECS and the Pacific Northwest National Laboratory (PNNL) provided the foundation for the U.S. Department of Energy's (DOE) Commercial Reference Building Models (CRBMs). From the extensive datasets,

15 commercial buildings types and 1 mid-rise apartment building complex were generalized (Deru, *et al.* 2011). CRBMs have been developed for three time periods: buildings constructed prior to 1980, buildings constructed after 1980 and new constructions. The DOE's energy simulation software, EnergyPlus, integrates the CRBMs and it generates energy models for commercial building types (Section 2.2, Phase 03).

Strategies: comparison of three rooftop strategies

Three rooftop strategies will be assessed: black synthetic rubber membrane, green roof agriculture and hydroponic greenhouse rooftop agriculture (Figure 12.1).

Evaluation: multifaceted criteria

Four performance indicators including building energy use, rooftop surface temperature, stormwater runoff and yield are being used to conduct the comparative analysis of the three roof strategies on the selected commercial building typologies.

GREEN ROOF AGRICULTURE

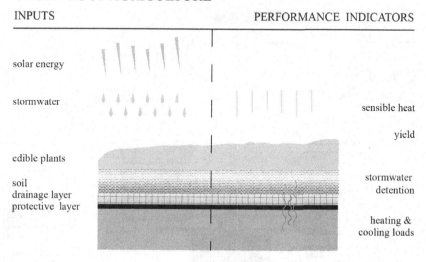

Figure 12.1 Two schematic diagrams depicting the environmental and system design inputs (left) and performance indicators (right) assessed for each of the retrofit strategies

Research phasing

This study uses a modeled case study approach. The existing urban fabric of Los Angeles provides the bases for the analysis, modeling and upscaling of the data. The research is divided into four phases and it explores three spatial scales – city, case study site and building (Figure 12.2).

Phase 01: City-scale spatial analysis

Phase 01 of the modeled case study analyzed the interconnections of climatic patterns, the physical environment and demographic profiles in the City of Los Angeles. The datasets were collected and generated from an array of interdisciplinary resources (Table 12.1). The data were spatially distributed using the 2010 census tracts. In ArcGIS 10.1 grouping analysis, a spatial analysis tool was used to evaluate multiple data inputs and to categorize them into groups with similar values (ArcGIS 2013).

Climate-related data for the grouping analysis were generated with point data downloaded from 12 Los Angeles County weather stations. The Spline tool was used to interpolate the point data and then it was joined to the census tracts' attribute table using the Zonal Statistic As Table tool. Multi-year averages (2003–2012)

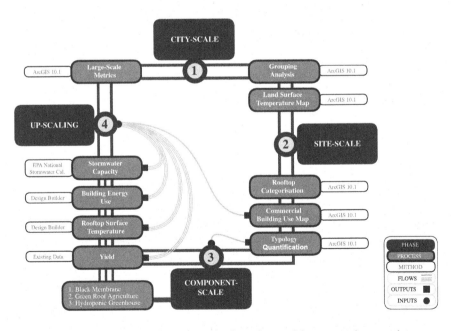

Figure 12.2 Schematic diagram depicting the four phases of the case study research process

Table 12.1 Set of interdisciplinary datasets used in the city-scale analysis phase

Category	Source	Datasets
Climate	NOAA Data Tools	Monthly Precipitation, Monthly Extreme Maximum Temperature, Monthly Cooling Days
Demographics	US Census: Fact Finder	Total Population, Participants in SNAP, Median Household Income
Physical	LA County GIS Data	Soil Map, Zoning, Building Footprints, Land Use, Sub-watersheds, Service Planning Areas, City Boundaries, Parcels, Roads, Public Transportation Lines, 2010 Census Tracts
Physical	USGS Earth Explorer	Landsat 7, High Resolution Orthoimagery
Physical	LA County Assessors	Local Roll (Building Use)
Physical	Soil Map & Soil Runoff Coefficient Data for LA	% Census Tract (CT) with 0.7 to 1.0 Runoff Coefficient
Physical	Zoning Map & CT	% CT with Commercial & Manufacturing Zoning
Physical	Zoning Map & Proportion Impervious	% CT with 80% to 100% Impervious Surface
Demographics	Food Access Research Atlas	Low Income & Low Access Measured at ½
Demographics	2011 LA County Health Survey	% Adult Obesity, % Adults Eating 5 Servings of Fruits & Vegetables Per Day, % Diabetes
Demographics	Cal EnviroScreen 2.0	Particulate Matter

were generated for monthly precipitation, extreme maximum temperature and the number of cooling days (NCDC 2014). Cooling days is a method to correlate daily temperature to the demand for energy to cool buildings (NWS 2014).

Food-related indicators from a 2013 Los Angeles County Department of Public Health publication were integrated in the grouping analysis (LA County DPH 2013). The data were downscaled from Service Planning Areas to census tracts using the tabulate intersection tool. Downscaling the datasets results in the approximation of weighted values in the census tracts for the percentage of obese adults, the percentage of adults that consume five or more servings of fruits or vegetables per day and the percentage of low-income, food-insecure households.

The datasets integrated into the grouping analysis contained parameters relevant to the four performance indicators (Table 12.2). However, only one temperature-related analysis field, the average maximum temperature in August, was included. Therefore to further analyze heat-related vulnerabilities a land surface temperature (LST) map for August 14, 2009 was generated from a LANDSAT 7.0 image (USGS 2014). Walawender, *et al.* (2012) have developed a thermal remote sensing tool for ArcGIS, LANDSAT TRS. This important tool was used to generate the LST map of Los Angeles used in the research.

Table 12.2 The set of analysis fields that was included in the final grouping analysis

Stormwater	Food	Temperature	Building energy use
Percentage of census tract with 0.7 to 1.0 runoff coefficient based on soil type	Percentage of adults consuming five or more servings of fruits & vegetables per day	Average extreme maximum temperature in August (2003–2012)	Percentage of census tract with commercial & manufacturing zoning
Percentage of census tract with 80% to 100% impervious surface	Percentage of households with incomes, 300% FPL who are food insecure		Annual average for total cooling days (2003–2012)
Annual average for total precipitation (2003–2012)	Percentage of adults who are obese BMI ≥ 30.0		

Phase 02: Case study site-scale assessment

Phase 02 assessed the existing building fabric within the modeled case study boundaries. In this phase the fundamental spatial dataset was the building footprints embedded with the building use codes. Google Earth's Street View supported the categorization of building uses when clarification was needed. Using a high resolution orthoimagery, qualitative analysis was conducted to verify that the site was comprised predominantly of flat roofs (USGS 2014).

The extensive array of municipal building codes in the study area was categorized in accordance with the Commercial Reference Building Models (Section 2.1). The GIS statistics tool in the attribute table was used to calculate the mean rooftop area, the mean building height, mean building area and the total rooftop area. The site-scale assessment provides the framework for scaling up the simulation data from Phase 03.

Phase 03: Component-scale simulation methods for the performance indicators

Phase 03 of the modeled case study seeks to quantify the four performance indicators of the rooftop strategies using the following methods.

Energy use and rooftop surface temperature: DesignBuilder

DesignBuilder's accessible platform is driven by the DOE's EnergyPlus software and it integrates 3D modeling and enhanced simulation capabilities (DesignBuilder 2014). It utilizes local climate data to model heating and cooling flow and other dynamic building properties (Sailor 2008). Output reports include data on building energy use and outside surface temperature.

The "pre-1980" CRBMs will be used in the energy analysis models since the majority of the buildings in the case study were constructed prior to 1980. A base 3D model for each compatible building typology will be modeled according to the mean building size, rooftop area and height parameters determined in Phase 02. Each rooftop strategies will be simulated, analyzed and compared.

EnergyPlus contains an Ecoroof module, which evaluates the impact green roofs have on the thermal load of an existing or a new building. The Ecoroof module contains many customizable parameters; four critical properties that significantly impact the performance of a green roof are soil depth, soil moisture, leaf area index (LAI) and irrigation schedules. The whole building analysis outputs from EnergyPlus include sensible heat transfers from soil and vegetation on the roof, stormwater runoff depth and heating and cooling loads in the building.

Stormwater: EPA national stormwater tool

The United States Department of the Environment (EPA) designed the National Stormwater Calculator (NSC) to calculate runoff from small-scale sites (Rossman 2014). NSC integrates historic precipitation data from the National Weather Services' (NWS) extensive network of rain gauges as well as future climate change scenarios from the World Climate Research Programme. For each rooftop strategy, a set of calculations is being generated to simulate their potential performance with historic precipitation averages and two regional near-term (2020–2049) climate change scenarios: one dry and one wet.

The model precipitation data references the NWS downtown LA/USC rain gauge, which had an average annual rainfall (1970–2006) of 39.19 cm (15.43"). For the same gauge the evaporation rate is 0.56 cm (0.22")/day. The BM baseline simulation assumes that 100 percent of the existing site, a rooftop, is impervious. The GRA strategy utilizes the low impact development controls in the NSC. The GRA strategy is being calculated with a 25.4 cm (10") soil depth, which is an estimated average that account for height variations in the landscape form used in production. It is assumed that the roofs in the case study area have a 2 percent slope. Agriculture is not included as a land cover type in the tool therefore the GRA strategy is being simulated with 50 percent meadow and 50 percent lawn land types to reflect variations in crop rotations. The method to calculate runoff from the HGRA strategy is still under consideration.

Preliminary results from the stormwater management potential of the green roof agriculture are significant when compared with the baseline black membrane strategy. On a 0.4 ha (1 acre) roof, simulated with local historic precipitation data, initial findings indicate that with the GRA strategy 55.5 percent of annual wet days could be detained as compared to only 12.5 percent with BM strategy. The largest rainfall event that could be detained without runoff would be 2.5 cm (1.06") for the GRA, while the BM strategy would only be 0.48 cm (0.19").

Yield: Analysis of existing data

Total yield potential will be estimated by averaging yield data from operational rooftop farms. Data is predominately being collected through in-person interviews and literature reviews. Currently, Los Angeles does not have any large-scale GRA or HGRA rooftop farms. Therefore, yield metrics from ground-level urban agriculture projects in the region will be collected. For GRA the rooftop yield data from locations outside the Southern California region will need to be adjusted to reflect the year round cultivation that is possible in Los Angeles. The three types of produce included in the study will be determined after data collection has been completed.

Phase 04: Upscaling the component data

In Phase 04 a set of large-scale performance metrics will be generated for each of the rooftop strategies (Table 12.3). For the GRA and HGRA strategies 10 percent of the total compatible rooftop area will be removed from the retrofit area to account for loss of space due to rooftop walkways, building equipment and other productive landscape needs. Building energy use per typology will be scaled up relative to the composition of building types in the modeled case study area. To calculate total stormwater runoff volume an area unit for each strategy will be determined and multiplied by the appropriate total area. For each type of produce, the yield potential per area unit will be scaled-up by the total growing area.

The anticipated datasets for each performance indicator are outlined in Table 12.3. Equivalencies will be estimated to make the datasets more meaningful and tangible to urban planning processes.

Table 12.3 Performance indicator paired with the scaling-up method

Analysis field	Overall: R2 alue	Group 07: Mean
% adult consumption of fruits & vegetables 5x / day	0.98	16.8%
% adult obesity	0.97	20.6%
% food insecurity	0.95	32.5%
Max. temperature extreme in August, average °f	0.75	92.3
Total number of cooling days	0.73	1064
% impervious based on zoning type	0.68	59.3%
Average annual precipitation (inches)	0.58	14.5
% tract with soil runoff coefficient from 0.7 to 1.0	0.53	84.1
% manufacturing and commercial zoning	0.45	38.1%

Results: Phases 01 and 02

Phase 01: Delineation of the modeled case study area

The objective of the city-scale spatial analysis (Phase 01) was to detect patterns in climatic, physical and demographic factors in the City of Los Angeles to understand where the adaptation strategy could be the most meaningful and effective at supporting multi-faceted urban resilience in the near-term.

The results from the grouping analysis informed the selection of the model case study site. To determine the group of census tracts that would be the most suitable for this research two values from the grouping analysis report were analyzed: (1) the $R2$ value in the overall variable statistics and (2) the mean value in the group-focused summaries (Figure 12.3). The $R2$ values were used to determine which analysis fields would be included in the final grouping analysis. High $R2$ values signal that an analysis field is significant in separating the groups (ArcGIS 2013). In the final grouping analysis all of the analysis fields had high $R2$ values, indicating that each field was important in grouping the census tracts. The mean values in the report were derived from the original weighted values in the attribute table for each analysis field.

Commercial buildings are the fundamental building block of this planned adaptation research. Therefore it was essential to find an area of the city with a high percentage of commercial and manufacturing zoning. Group 07 had the highest mean value for these types of zoning (38 percent), satisfying the principle site selection criteria. Further analysis was conducted to evaluate if Group 07's risk profile had significant social and climatic vulnerabilities relevant to this study.

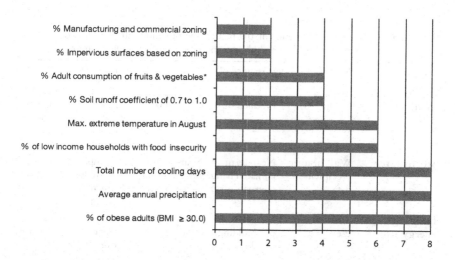

Figure 12.3 The overall $R2$ values and the mean values of Group 07 for each analysis field compared with the other groups

To evaluate the risk profile of Group 07 in relation to the other groups, the mean values from the Grouping Analysis Report were ranked (Table 12.4). Group 07 had the highest percentage of impervious surfaces and the second highest area of soil with a high runoff coefficient. It had the second lowest percentage of adults consuming the recommended five servings of fruits and vegetables per day. Furthermore, it had the third highest levels of low-income, food insecure households and maximum extreme temperatures in August. The total precipitation and total cooling days were average for the City of Los Angeles. Cumulatively the census tracts included in Group 07 had a slightly lower than average ranking for the percentage of adults with obesity, although downscaling the data could have distorted this analysis field.

The grouping analysis ranking process revealed elevated stormwater-related, heat-related and food-related vulnerabilities in Group 07. Therefore, exploring the potential impacts from retrofitting existing commercial roofs with productive landscapes within Group 07 aligns with the resiliency objectives of this research. The multi-faceted rooftop retrofit strategy could meaningfully reduce several existing vulnerabilities in these neighborhoods in Los Angeles.

The spatial distribution of the eight groups generated during the grouping analysis is mapped in Figure 12.4. Group 07 is distributed in dispersed clusters in the northern and southern regions of Los Angeles, with the densest aggregation of census tracts around the downtown area. Due to the expanse of commercial buildings in this area it was chosen as the region of interest for the research on adapting existing commercial rooftops.

Table 12.4 Based on an annual rainfall, the percentage of runoff, infiltration and evaporation with black membrane and green roof agriculture strategies

Indicator	*Datasets per rooftop strategy*			*Equivalency method (in progress)*
Yield	Produce 01	Produce 02	Produce 03	Nutritional value as a % of recommended intake for Healthy People 2020
Stormwater	With historic precipitation data	Near-term climate change scenario (dry)	Near-term climate change scenario (wet)	Compared to conventional stormwater infrastructure costs
Building energy use	Total energy use	Energy costs per building typology	Emissions per building typology	GHG emissions equated to number of cars
Rooftop surface temperature	Summer roof surface temperature			TBD (a per area metric for air temperature change)

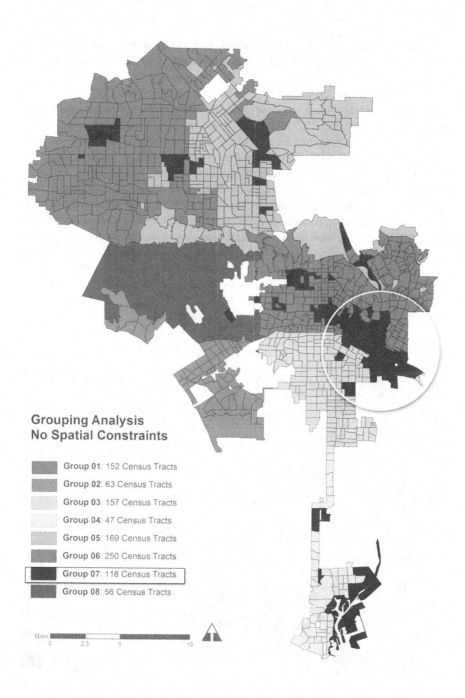

Grouping Analysis
No Spatial Constraints

	Group 01: 152 Census Tracts
	Group 02: 63 Census Tracts
	Group 03: 157 Census Tracts
	Group 04: 47 Census Tracts
	Group 05: 169 Census Tracts
	Group 06: 250 Census Tracts
	Group 07: 118 Census Tracts
	Group 08: 56 Census Tracts

Miles
0   2.5   5   10

Figure 12.4 Grouping analysis mapping for Los Angeles

Note: The focal group is Group 07. The circle delineates the region of interest in the downtown area.

Finally, the exact boundary of the modeled case study area was informed by the LST map, which allowed for a finer grain analysis of heat-related vulnerabilities in the City of Los Angeles (Figure 12.5). On August 14, 2009 the maximum air temperature reached 28°C (82.4°F) (Weather Underground 2014). The land surface temperatures in and southeast of the downtown area reached the maximum surface temperature (37.2°C/99.0°F) for that day. In this area there was a 9.2°C (16.6°F) difference between the air and the land temperature. Although the LST map represents a single day, it clearly illustrates existing and future heat-related vulnerabilities for high-temperature days in Los Angeles.

In conclusion, the outcome of the city-scale spatial analysis was to identify a site with a high percentage of commercial-manufacturing zoning with predominately flat roof buildings coupled with elevated risks in food, heat and stormwater. Results from GIS's grouping analysis and the LST mapping were the predominate methods for delineating the case study site. The city-scale spatial analysis provided the statistical data needed to validate the exploration of large-scale adaptation of existing rooftops with productive landscapes in this area of Los Angeles.

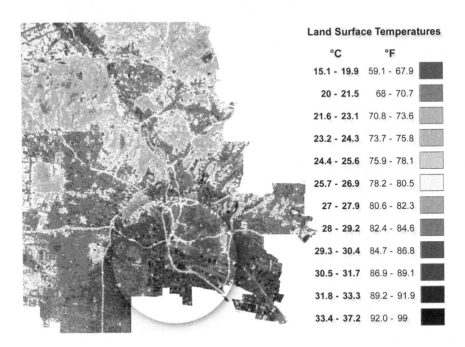

Figure 12.5 Land surface temperature map for August 14, 2009 generated with the Landsat TRS tool

Note: The minimum air temperatures was 22°C (71.6°F) and the maximum temperature was 28°C (82.4°F). The modeled case study area is within the circle.

Phase 02: Categorizing the rooftop landscape

This research focuses on evaluating the potential impacts from adapting existing commercial buildings in the City of Los Angeles with a targeted green infrastructure strategy. The objectives of Phase 02 were to catalogue all of the building typologies in the case study area, to determine which uses were appropriate for the productive rooftops and to calculate essential mean averages for each compatible building typology to provide modeling parameters for the building energy use simulations.

Seven appropriate building use typologies were identified: educational facility, residential condominiums, stand-alone retail, industrial warehouse, office buildings, restaurants and supermarkets (Figure 12.6). Compatibility was based on existing use as well as the potential to foster a cooperative relationship with a productive rooftop. Excluded building uses included heavy industrial facilities, service stations, police and fire stations, parking lots and vacant land.

The total surface area from the compatible building rooftops represented 37 percent, 500 ha (1,236 acres), of the total area within the site. Accounting for a 10 percent loss in total area for the GRA and the HGRA strategies the total area that will be modeled with productive landscape in this study is 450 ha (1,112 acres).

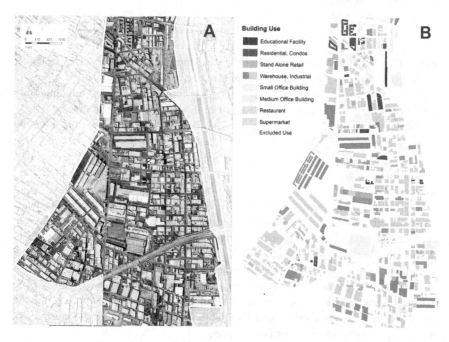

Building Use

- Educational Facility
- Residential, Condos
- Stand Alone Retail
- Warehouse, Industrial
- Small Office Building
- Medium Office Building
- Restaurant
- Supermarket
- Excluded Use

Figure 12.6 a) USGS high resolution orthoimagery of a portion of the case study area; b) The same area mapped with building use symbology

This is roughly two times the size of Olympic Park in London or the approximate size of Golden Gate Park in San Francisco. In the landscape of buildings, warehouses with rooftops greater than or equal to 4645 m² (50,000 ft²) dominate (55 percent), followed by warehouses with rooftops less than 4645 m² (41 percent) and office buildings (3 percent) comprise a distant third.

The mean building age, height, area and rooftop surface area were calculated to begin to comprehend the physical characteristics of the commercial building stock. Based on available data the average year of construction for the buildings is 1959. The building fabric is comprised predominately of low-rise buildings with mean heights ranging from 4.9 m to 12.2 m (16' to 40'). The building types with the greatest mean rooftop surface area are large-scale warehouses (30,236 m²/99,198 ft²), stand-alone retail (12,281 m²/40,292 ft²) and condominiums (1,380 m²/14,860 ft²).

Discussion

En masse, the rooftops of existing buildings are an underutilized, inefficient urban landscape. Adapting rooftops with green roofs is an opportunity to blur the boundaries of the bifurcated urban and nonurban model of development. The first step is to value rooftops as a network of dispersed planes that collectively form the building blocks of a multi-functional urban system. Large-scale, commercial rooftop farming is an emerging production model and it has the potential to be a new, high-performing system integrated into existing urban infrastructure.

The primary objective of this research is to elevate urban agriculture – physically and conceptually. Densification of cities has intensified two issues at the nexus of urban agriculture and urban planning: (1) the competition for the allocation of space and (2) linkages in human health, environmental health and the design of urban environments. Elevating food production onto underutilized rooftops of existing commercial buildings alleviates space limitations while simultaneously providing an array of social and ecological co-benefits to urban communities.

Within the boundaries of this Los Angeles-based case study, the compatible use rooftops aggregates to 37 percent of the total area. In a dense urban fabric that statistic represents an immense opportunity to integrate the production of nutritious food, improve urban hydrology, expand the use of heat-mitigating plants and reduce GHG emissions from existing buildings.

This modeled case study is bolstered by significant research on productive landscapes in urban areas, for example, the extensive research program, Five Borough Farm, dedicated to bolstering urban agriculture production and planning processes in the New York City area (Cohen et al. 2012). Proksch's comprehensive research specifically on rooftop farming demonstrates the environmental benefits – stormwater management, energy-savings, cooling and carbon sequestration – from existing rooftop farms in various cities (Proksch 2014). This research builds on precedent like these and other interdisciplinary research to explore the potential of rooftop farming in the City of Los Angeles.

The diet–health linkage provides a particularly compelling case for integrating the production of healthy food on the rooftops of existing commercial buildings in Los Angeles. City resilience is fundamentally rooted in the health of its inhabitants and the urban environment. Research on rooftop farming at an urban scale supports the development of a robust, equitable urban food system in the near term and in the long term it is an opportunity to improve the overall resilience of the City of Los Angeles.

References

Alexandri, E. and P. Jones (2006) Temperature decreases in an urban canyon due to green walls and green roofs in diverse climates. *Building and Environment* (Science Direct): 480–493.

ArcGIS (2013) How Grouping Analysis Works. *ArcGIS Help 10.1.* 4 18. http://resources.arcgis.com/en/help/main/10.1/.

Bloom, A.J., M. Burger, B.A. Kimball, and P.J. Pinter Jr (2014) Nitrate assimilation is inhibited by elevated CO2 in field-grown wheat. *Nature Climate Change*: 477–480.

Brooklyn Grange (2015) *About.* http://brooklyngrangefarm.com/about/.

Carter, T., and C. Butler (2008) Ecological impacts of replacing traditional roofs with green roofs in two urban areas. *Cities and the Environment* 1 2: 1–17.

Castleton, H.F., V. Stovin, S.B.M. Beck, and J.B. Davison (2010) Green roofs; building energy savings and the potenial for retrofit. *Energy and Buildings*: 1582–1591.

Cohen, N., Reynolds, K., and Sanghvi, R. (2012). *Five Borough Farm: Seeding the Future of Urban Agriculture.* Design Trust for Public Space.

Custot, J., M. Dubbeling, A. Getz-Escudero, J. Padgham, R. Tuts, and S. Wabbes (2012) Resilient Food Systems for Resilient Cities. In *Resilient Cities 2: Cities and Adaptation to Climate Change*, edited by Konrad Otto-Zimmermann, 125–138. Bonn: Springer.

Deru, M., K. Field, D. Studer, K. Benne, B. Griffith, and P. Torcellini (2011) *U.S. Department of Energy Commercial Reference Building Models of the National Building Stock.* Technical, National Renewable Energy Laboratory, Golden: NREL.

DesignBuilder (2014) *DesignBuilder Software Product Overview.* www.designbuilder.co.uk/.

EPDM Roofing Association (2014) *What is EPDM.* www.epdmroofs.org/what-is-epdm (accessed 09 14, 2014).

Gill, S.E., J.F. Handley, A.R. Ennos, and S. Pauleit (2007) Adapting cities for climate change: The role of the green infrastructure. *Built Environment* 33: 115–133.

Gotham Greens (2015) http://gothamgreens.com/our-farm/.

Hassan, A. (2010) EGYPT: Fears of a food crisis after Russia's wheat export ban. *Los Angeles Times*, 8 August.

IPCC (2012) *Managing the Risks of Extreme Events and Disasters to Advance Climate Change Adaptation.* Assessment, Cambridge: Intergovernmental Panel on Climate Change.

LA County DPH (2013) *Key Indicators of Health by Service Planning Area.* Health, LA County DPH, Los Angeles: LA County DPH.

Larsen, L., *et al.* (2011) *Green Building and Climate Resilience: Understanding Impacts and Preparing for Changing Conditions.* Urban and Regional Planning Program, U.S. Green Building Council, Ann Arbor: University of Michigan.

Levine, M., *et al.* (2007) *Residential and commerical buildings.* Fourth Assessment Reprot of the Intergovernmental Panel on Climate Change, Working Group III, Cambridge: Cambridge University Press.

Long, S.P., E.A. Ainsworth, A.D.B. Leakey, J. Nosberger, and D.R. Ort (2006) Food for thought: Lower-than-expercted crop yields simulations with rising CO2 concentrations. *Science* 312: 1918–1921.

Lufa Farms (2015) *Laval*. http://lufa.com/en/our-farms.html.

Moberg, F. and S.H. Simonsen (2014) *What is Resilience? An introduction to social-ecological research*. Stockholm: Stockholm Resilience Center, 3.

NCDC (2014) *National Climate Data Center – Find Station*. www.ncdc.noaa.gov/cdo-web/datatools/findstation.

NIH (2011) *What is Metabolic Syndrome*. www.nhlbi.nih.gov/health/health-topics/topics/ms.

NWS (2014) *National Weather Service – Degree Days*. www.erh.noaa.gov/cle/climate/info/degreedays.html.

NWS Forecast Office (2015) *NOWdata – NOAA online Weather Data*. http://w2.weather.gov/climate/xmacis.php?wfo=lox.

Oberndorfer, E., *et al.* (2007) Green roofs as urban ecosystems: Ecological structures, functions, and services. *BioScience* 57 10: 823–833.

ØsterGro (2015) www.kobenhavnergron.dk/place/ostergro/?lang=en.

Pearson, L.J., L. Pearson, and C.J. Pearson (2011) Sustainable urban agriculture: stocktake and opportunities, *International Jounal of Agricultural Sustainability* 8: 7–19.

Perry, M.L., O.F. Canzaiani, J.P. Palutikof, P.J. van der Liden, and C.E. Hanson (2007) *Climate Change 2007: Working Group II: Impacts, Adaptation and Vulnerability*. Assessment, Working Group II, IPCC, Cambridge: Cambridge University Press.

Proksch, G. (2014) Urban rooftops as productive resources rooftop farming versus conventional green roofs. *ARCC*: 497–509.

Rossman, L.A. (2014) *National Stormwater Calcuator User's Guide*. User's Guide, Office of Research and Development, United States Environmental Protection Agency, Cincinnati: US EPA.

Sailor, D.J. (2008) A green roof model for building energy simulation programs. *Energy and Buildings*, 02: 1466–1478.

Sailor, D.J., T.B. Elley, and M. Gibson (2011) Exploring the building energy impacts of green roof design decicions – a modeling study of buildings in four distinct climates. *Journal of Building Physics* 35 4: 372–391.

Simmons, M.T., B. Gardiner, S. Windhager, and J. Tinsley (2008) Green roofs are not created equal: the hydrologic and thermal performance of six different extensive green roofs and reflective and non-reflective roofs in a sub-tropical climate. *Urban Ecosystem*: 339–348.

Spala, A., H.S. Bagiorgas, M.N. Assimakopoulos, J. Kalavrouziotis, D. Matthopoulos, and G. Mihalakakou (2008) One the green roof system. Selection, state of the art and energy potential investigation of a system installed in an office building in Athens, Greece. *Renewable Energy* 33: 173–177.

U.S. DOE (2010) Energy efficiency & renewable energy. U.S. Department of Energy. http://buildingsdatabook.eren.doe.gov/.

Union of Concerned Scientists (2012) *Preparing for Climate Change Impacts in Los Angeles*. Fact Sheet, Berkeley: Union of Concerned Scientists.

United Nations (2012) *World Urbanization Prospects: The 2011 Revisions*. Highlights, Department of Economics and Social Affairs / Population Division, United Nations.

USGS (2014) *EarthExplorer*. http://earthexplorer.usgs.gov/.

Walawender, J.P., M.J. Hajto, and P. Iwaniuk (2012) A new ArcGIS toolset for automated mapping of land surface temperature with the use of LANDSAT satellite data. *Proc. IEEE International Geoscience and Remote Sensing Symposium (IGARSS)*. Munich: 4371–4374.

Weather Underground (2014) *Weather History*. www.wunderground.com/.

Whitford, V., A.R. Ennos, and J.F. Handley (2001) City form and natural process: Indicators for the ecological performance of urban areas and their application to Mersey, UK. *Landscape and Urban Planning*: 91–103.

13 The role of place, community and values in contemporary Israel's urban agriculture

Tal Alon-Mozes

Introduction

Friday morning is seemingly a busy time for the community of gardeners at A Garden for the Resident, a municipal urban agriculture allotment located not far from the center of Rishon LeZion, Israel's fourth largest city. Dozens of people wander around their plots, griping about the broken water pipe, heat, humidity, mosquitoes and the current political situation. By the end of the summer there is very little to do, so they pass the time in the shade under lavish vine pergolas drinking black coffee and eating fresh grapes, before returning home to attend to weekend chores. A kilometer westward, in the neighborhood of Ramat Eliyahu (also belonging to Rishon LeZion), the Ethiopian community's garden is closed and empty; activity is restricted solely to Monday and Tuesday afternoons.

These two urban agricultural sites represent the growing interest of individuals, communities, grassroots organizations, municipalities and state institutions in the establishment of edible gardens throughout Israel. In a recent survey, the organization *Community Gardens in Israel* identified more than 300 community-based gardens, all involved to a certain degree in growing edible produce (http://israel-gardens.blogspot.co.il/2009/06/blog-post.html). These gardens include allotments, community gardens within public institutions, community gardens on private or public lands, therapeutic gardens, urban community farms, municipal nature sites and community forests.

Compared to results of a survey conducted in 2004, in which we located approximately two dozen urban agricultural sites throughout the country (Alon-Mozes & Amdur, 2005), the tenfold increase reflects various social, economic and cultural tendencies that have developed over the past ten years in Israel. Among affluent communities, these include the adoption of a "green agenda", an increase in the consumption of healthy foods and an inclination to belong to a community. Among the poorest communities (especially among immigrants), it reflects a decrease in food security and nostalgia for the community's agricultural past. For both communities, it reflects the adoption of urban agriculture models from around the world.

This chapter focuses on two case studies of contemporary urban agriculture in Israel, representing two extremes of the current practice. While A Garden for the

Resident reflects a grassroots initiative of an affluent community, the Ethiopian Community Garden represents a municipal initiative to support the poorest community of new immigrants in the same city. While the projects differ in terms of scale, organization and labor practices, they are similar in terms of produce and, to some extent, in the manner in which they benefit their participants.

The chapter consists of three parts: the first discusses the historical roots of urban agriculture in Israel; the second surveys the contemporary state of affairs; and the third elaborates on the two case studies. The research is based on on-site observations and interviews with gardeners and stakeholders. The results are compared with a previous survey that we conducted in Rishon LeZion between the winter of 2004 and the spring of 2005.

Historical background

Urban agriculture was a common practice among Zionist settlers in Palestine since the early twentieth century. In hundreds of small urban farms located in the peripheral outskirts of dense urban centers such as Tel Aviv and Haifa, people grew fruits and vegetables and some also raised chickens and milk cows on plots that diverged in size from less than 250 sqm to 200 sqm. The small urban farms drew on multiple precursors, reflecting the origins of the members of Zionist settling organizations, their political views and the influence of late nineteenth- and early twentieth-century ideas of rural/urban settlements in Germany and in other European countries. Among the social movements that influenced the creation of small urban farms in Palestine were the inner colonization movement, the supplementary homestead movement and the allotment movement (Alon-Mozes, 2015).

In Palestine, the small urban farms were established by various settling organizations, mainly those connected to the Labor Party. They were supported by local committees, and especially by women's organizations that helped the gardeners with planning schemes, supply of seeds and constant guidance, including how to prepare tasty meals from the unfamiliar vegetables.

In addition to serving the families with fresh produce and helping overcome periods of food insecurity, the establishment of these farms reflected the agricultural ethos dominant among the newcomers, who aspired to return to their ancestral homeland as farmers, to cultivate its land and thereby to contribute to the strengthening of ties between the Land of Israel and the Jewish people after thousands of years in exile. In lieu of insufficient land resources and the financial support necessary for developing rural areas, thousands of immigrants settled in urban centers. For many, this was perceived as an intermediate phase, as they held on to their dream of cultivating Israel's rural regions. Meanwhile, the small urban farms strengthened communities' cohesiveness, with people gathering around these gardens for festivals and competitions. Most of the gardens were the domain of the women, who remained at home while their husbands earned the family living in the nearby towns. Overcoming the difficulties and serving the family with fresh produce contributed to these women's sense of self-worth and accomplishment (Alon-Mozes, 2007).

Figure 13.1 Small urban farm in Kiryat Avoda, mid-1930s
Source: Holon Archive.

In pre-state Israel, urban agriculture was also practiced in women-workers urban farms – *Mishkei Poalot*. Located in several cities and townships, these farms not only trained women in agricultural practices, but also enabled them to sell their produce – vegetables, flowers, eggs and milk, to city dwellers (Katz & Neuman, 1996).

With the founding of the State of Israel, hundreds of small urban farms were established in new towns, all over the country, as part of the state's first master plan in the early 1950s. Although promoted by state authorities on radio programs and in a monthly journal, and despite the devoted guidance of agricultural advisors, the small urban farm project slowly waned for various reasons. Among them were the rapid development of the agricultural sector, which made home-grown produce unnecessary; the demand for land for housing; and the difficulty of maintaining a garden in the harsh environmental conditions – the hot climate, infertile soil and frequent sandstorms, mainly in southern Israel. However, the most important contributor to this demise was the human factor. While early urban gardeners were European settlers who had acquired urban farming practices in their countries of origin, the new gardeners, mostly immigrants of North African and Middle Eastern descent, were less motivated to become farmers.

In the years since, this tendency has been partly reversed, and urban agriculture is currently gaining popularity in various Israeli communities.

Contemporary urban agriculture in Israel

In late 2004 we identified a change in public opinion in favor of urban agriculture and conducted a survey on the state of art and feasibility of advancing urban agriculture in Israel. It showed that in twenty-first century Israel, the domestic cultivation of conventional or even organic food produce bears no economic value. However, there are other motivations behind the growing popularity of the practice. In a country in which 91 percent of the population live in urban settlements and only 2 percent earns its living from agriculture, practicing urban agriculture represents not only a sense of connectedness to the land, but also a growing awareness to healthier nutrition and an aspiration for a more sustainable way of life. Two groups within Israeli society were identified as willing to take part in urban agriculture: affluent populations that perceive it as a leisure activity; and poor populations, or those with special needs, who benefit from the nutritional and therapeutic values associated with growing herbs, fruits and vegetables.

The survey identified five dominant obstacles to implementing the project: land availability and ownership, scarcity of water and high water prices, planning, organization and public awareness. It suggested a list of organizations on the state or municipal level to manage the project in collaboration with local grassroots organizations, and creative means to overcome the water problem by using runoff water, grey water, spring water or subsidized water, as common in the agricultural sector. The survey emphasized the need for intensive support of professionals.

Since 2004, urban agriculture in Israel grew by more than tenfold. According to the Organization of Community Gardens in Israel (http://israel-gardens.blogspot.co.il/), which is a non-profit organization with a vibrant website, in 2014 there were more than 300 community gardens throughout the country. They diverge in their goals, scope of practice and target population: 88 of the 330 gardens (27 percent) are located around Israeli metropolitan centers of Tel Aviv, Jerusalem and Beer Sheba; 62 gardens (19 percent) are located in the periphery; 19 gardens were established for the benefit of the Ethiopian community, and 87 for other mixed groups of immigrants.

According to the survey, the hundreds of gardens can be divided into eight typologies, as described in Table 13.1.

A contemporary follow up on the 2014 list reveals the temporariness of many of these gardens, which change their character over the years. In general, it seems that while community gardens and urban agriculture are becoming more popular nowadays, the appropriate models for Israel are still emerging.

The rest of the chapter discusses two extremes of these models, both located in central Israel.

A Garden for the Resident

A Garden for the Resident, which represents Israel's version of an allotment, is one of the most prominent contemporary urban agriculture projects in Israel. The garden was established in 2005 by the municipality of Rishon LeZion, one of the

Table 13.1 Types of Israel's urban agricultural gardens

Garden typology	Community definition	Land ownership	Goals	Mode of activity	Total number
Community gardens	Open to all/specific community	Public	Various: mainly social but also environmental, and educational	Mainly communal, but sometimes people cultivate personal beds. Social structure differs from site to site, but usually a core group of activists leads the garden.	178
Allotments	Open to those who pay. Waitlisted	Public, private recreation	Food production,	Each plot holder works his plot	6
Community gardens in public institutions	Specific community which is related to the institution	Public, or belongs to the institution	Mainly social, strengthening community cohesiveness or institutional goals	Communal activity	59
Urban community farms	Everyone	Public/private	Agricultural/educational, community	Communal activity which integrates gardening with other social and educational activities	2
Therapeutic gardens	Specific community	Public/private	Therapeutic	Various models: individual alongside communal work	46
Private land under community care (for example; common ground of tenements)	Open to all/specific community	Private	Recreation, social gathering	Communal activities organized by residents or by NGO groups	20
Unique gardens/ municipal nature site	Open to all/specific community	Public	Nature care	Communal activity, including gardening	4
Community forest	Open to all		Forest care	Communal activity, including gardening	15

densest cities in central Israel, as the fulfillment of a pledge made during the 2004 municipal elections. The garden is adjacent to a highly populated neighborhood, and is located on land where construction is prohibited due to an on-site high voltage electricity line and underground gas pipe. The garden's terrain, which is particularly fertile, was divided into plots of 100–150 sqm each, all delineated by a low metal fence and cultivated by individual gardeners.

According to our 2004–2005 survey, most Rishon LeZion gardeners were married middle and upper-middle class males, with three-quarters more than 40 years old. Two-thirds of the gardeners were either born in Israel or immigrated as children. These gardeners spend many hours on their plots: almost half of them reported that they visit the site several times a week, while a third said that they visit daily. Half of the gardeners spend more than two hours on each visit.

As mentioned above, in contemporary Israel the economic value of practicing urban agriculture is negligible. Most of the gardeners even reported that the cost of plot cultivation exceeded its production value. However, we found that for most Rishon LeZion gardeners, plot cultivation was associated with personal or ideological values. For some, it was the significance they attributed to organic farming; while for others it was the embodiment of the experience of cultivating the land, of creating an intimate connection with the soil, which one gardener described as "giving birth". Other gardeners focused on a strong emotional attachment to the site; at times it was associated with nostalgic memories of their childhood (Alon-Mozes & Amdur, 2010).

Figure 13.2 Garden for the Resident (Rishon LeZion, 2005)

A sense of community in the Garden for the Resident project was evident in three different spheres: the family, neighboring gardeners and the local general public. Many gardeners mentioned that they share gardening work with family members, mainly with spouses and children. Elderly gardeners noted that they enjoy working with their grandchildren, as well as with other gardeners who they had previously not been acquainted with. They emphasized mutual assistance and the sharing of tools, seeds, vegetables and flowers, as well as knowledge and expertise. They valued these friendships for their unique qualities.

During our recent visit, almost ten years after our first survey, we found that the garden is prospering. A 2 m wire fence with a locked gate surrounds most of the plots. Covered in vines, these fences represent a preference for privacy over the communal values of the past. The gardeners grow vegetables, flowers and fruit trees on two thirds of their plots, while the remainder consists of shady rest areas and small kitchens.

The project is well maintained and there are no vacant plots. There is a waiting list for plot rentals (at approximately $300 a year); and several members sold their rental rights at a relatively high price (ranging from twice to seven times the original annual rent rate). These increased rates are justified in terms of investments made on the sites, including the construction of pergolas, irrigation systems, paved

Figure 13.3 Garden for the Resident (Rishon LeZion, 2014)

areas and most of all, the nurturing, over time, of a productive garden. In the past nine years, the gardens gradually matured, as the gardeners' growing expertise resulted in a significant increase in crop variety and yield. Some gardeners experiment with unique imported species; one even produces wine. Participants share their success with friends, family and enthusiasts in Israel and abroad via social networks, on which they post photographs of their finest produce.

In 2010 the municipality installed individual water meters on each plot. As a result, gardeners were encouraged to use a computer regulated drip irrigation system to maximize irrigation efficiency. Other changes also reflect a shift in regulations. For instance, while a decade ago the municipality forbade planting trees, currently it permits planting non-aggressive species whose maximum growth height does not exceed their regulations. This has had a significant impact on the landscape, as these trees require little maintenance. In terms of the aging of the gardener community, we found that their children and grandchildren are less interested in gardening *per se*, and as several reported, they spend more time socializing with neighbor gardeners than cultivating their plots. Finally, in an attempt to increase ongoing municipal maintenance, the gardeners appointed a local committee to represent them in negotiations with local authorities.

The garden is a gated community within Rishon LeZion. Only key holders can enter the area, and currently several private patches are double locked. While children passing by are invited by gardeners to taste their produce, no official municipal activities take place in the garden. Located not far from the center of the fourth most populated city in Israel, the allocation of a hectare of public land for the benefit of a few raises questions regarding the right of the public at large to gain access to the public amenity.

A kilometer to the west, on the opposite side of the state highway that runs through the city, the Rishon LeZion municipality established two completely different community gardens solely for the city's Ethiopian community.

Ethiopian community and gardens

Community gardens for the Ethiopian community are a relatively new and interesting phenomenon in Israel, especially considering that the majority of its members arrived in Israel nearly 25 years ago.[1] At the end of 2012, the Ethiopian segment of the Israeli population comprised of 131,400 residents; approximately 84,600 were born in Ethiopia, while 46,800 were born in Israel. (www.cbs.gov.il/reader/newhodaot/hodaa_template_eng.html?hodaa=201311300). The Ethiopian community is considered poor.[2] Its members tend to live in close proximity to one another, inhabiting low-income neighborhoods in cities and towns situated all over Israel (www.iaej.co.il/?p=4853). Despite the considerable support of the Israeli establishment, this community has barely managed to integrate into Israeli society, and it faces significant economic and social difficulties. As in many other immigrant communities, its elders face the most severe challenges.

Prior to their immigration, many Ethiopian Jews were farmers. Many continued to practice agriculture when relocated to emigration camps in Ethiopia where they resided prior to their immigration. Upon arrival in Israel in the mid-1980s, they were settled in temporary immigrant camps, where they set up small gardens (Helphand, 2006). Those who arrived later were relocated to absorption centers in various towns, where many still reside. Their immigration and settlement in urban centers abruptly disconnected the community from its agricultural past and traditions, leaving it with no access to the countryside and to agricultural practices that could ease their absorption.[3]

The first Ethiopian community gardens in Israel were established in the late 2000s as initiatives of municipalities and public support organizations. Currently numbering more than 20, they lack centralized coordination although some organizations play a more dominant role in the initiative than others. Among these organizations are welfare NGOs established to assist the elderly population, others which support the Ethiopian community and to a lesser extent, environmental NGOs.

The JDC-ESHSL (Joint Distribution Committee), a philanthropic organization supporting diverse projects for the elderly, initiated a special communal gardening program for Ethiopians over 55 years of age. Community members permanently gather in these gardens, grow food, cook and eat together. The gardens also host community events for the general public (www.eshelnet.org.il/default.asp?catid= {20F46CAA-BC36-45CC-8EDD-5126752DE5FB}&details_type=1&itemid= {A12504EB-0446-4990-A741-D6B095F04A8A}).

Other organizations emphasize environmental awareness among city residents as their primary goal. Through its urban branches, SPNI (Society for the Protection of Nature in Israel), Israel's oldest and largest non-profit environmental organization, is responsible for the establishment of dozens of community gardens in cities throughout the country. As their activity is designated for the public at large, SPNI tends to include the Ethiopian community within existing gardens, in addition to supporting gardens intended solely for their behalf.

An organization that integrates both social and environmental goals is Earth's Promise (Shvuat HaAdama), a non-profit organization founded in Beer Sheba in 2007 by an American immigrant with a background in ecology. Since its establishment, the organization has inaugurated three community gardens adjacent to the town's absorption centers, and is currently in the process of developing an urban farm. Its mission emphasizes the "improvement and safeguard of the quality of Israel's environment by creating replicable grassroots models of sustainable urban development" (http://earthspromise.org/about/mission/). In practice, the majority of the projects serve the Ethiopian community. Other initiatives include the construction of a traditional Ethiopian mud house (*Godjo*) on one of the garden sites, where gardeners host a storytelling project; and since 2011, the annual festival of Ecothiopia, a cultural event in which hundreds of community members and guests gather for two days of workshops on ecology, ethnic dancing, singing and storytelling (http://shadama.org.il/en/).

In the gardens of Beer Sheba and of other Ethiopian communities, gardening is practiced collectively. Under the supervision of a community activist, gardeners gather once or twice a week to work and share their produce with community

Figure 13.4 Ethiopian Community Garden (Beer Sheba, 2013)

members. They mainly grow vegetables commonly used in the Ethiopian cuisine such as tomatoes, potatoes, peppers, eggplant, garlic, onion, mint, Ethiopian lettuce and a special cereal used for baking traditional Ethiopian bread. These gardens' economic value, according to the community activists, is very limited.

Lately, several new initiatives have aimed to increase these gardens' economic viability. Among these are two farms in central Israel where Ethiopians grow herbs for sale, in parallel with serving as an educational center for both the Ethiopian community and others who visit it for educational and recreational purposes (http://atachlit.wix.com/atachlit-eng).

The Ethiopian Community Garden in Rishon LeZion

Over 7,000 people of Ethiopian origin live in Rishon LeZion, mainly in the neighborhood of Ramat Eliyahu. In 2010 the municipality, in conjunction with the local absorption center, initiated the first community garden (today there are two) in order to provide occupational opportunities for Ethiopian elders.

The Ethiopian Community Garden is located on the outskirts of Ramat Eliyahu, which comprises three-story tenements densely populated by a majority of Ethiopian immigrants. Situated adjacent to a new synagogue, the garden covers an area of 200 sqm and is enclosed by a high wire fence. As there are only two key holders, opportunities to work in the garden are restricted to two afternoons a week.

Figure 13.5 Ethiopian Community Garden (Rishon LeZion, 2014)

On a Monday afternoon in mid-August 2014, approximately 20 men and 2 women gathered in the garden with small gardening tools. Though each was allotted a 10 sqm patch, they preferred working together. The patches are elevated by wooden boards and irrigated daily by a computerized drip system. Corn and red pepper are the garden's prevalent produce during the summer season. In addition, the gardeners grow special herbs whose seeds were imported from Ethiopia, as well as cereal plants used to make traditional Ethiopian bread. As opposed to the vegetable beds, these are few and mostly carry only symbolic meaning. The gardeners are aged 40 years and over, and their knowledge of Hebrew is limited. The two hours spent in the garden are devoted mainly to weeding, seeding corn, planting young seedlings in new spots, and watering the ground using a watering can, as most gardeners do not rely on the efficiency of the drip system. At seven o'clock in the evening they leave the garden with bunches of herbs; some visit the nearby synagogue for evening prayers.

The municipality and JDC-ESHEL support the garden. The on-site overseer is a member of the Ethiopian community. Within this poor and neglected neighborhood, the garden constitutes a lush, green oasis. However, the lavish atmosphere of the garden do not influence the surrounding areas and no plants find their way from the gated garden to the bare ground around the community dwellings.

Discussion

A distance of 1 km, an interstate highway and a significant social gap separate the Garden for the Resident and the Ethiopian Community Garden of Rishon LeZion. None of the Garden for the Resident members have ever visited the Ethiopian gardens and vice versa. In fact, most members in both groups do not know where the other's garden is located. Along the wide spectrum of communal gardening and urban agriculture projects, these two are located at the far ends. Table 13.2 reflects their differences.

Despite the substantial differences between these two environments, it seems that both are essentially motivated by a strong belief in the role of agricultural practices in promoting wellbeing. Both veteran Israelis and Ethiopian newcomers practice urban agriculture as a leisure activity rooted in nostalgia for past

Table 13.2 Comparison between Garden for the Resident and Ethiopian Community Garden

Characteristics	Garden for the Resident	Ethiopian Community Garden
Location	Attractive setting, not far from the city center. Municipal land, unavailable for development due to existing infrastructures. Within driving distance for most of the gardeners.	Attractive setting for the community. Municipal land. Within walking distance from the tenements.
Initiators	Bottom-up initiative of local residents. The project is managed by the Municipal Company of Rishon LeZion for Culture, Sport and Recreation Ltd.	Top-down initiative of the town's absorption center/ the municipality's welfare unit in collaboration with state-wide social and environmental organizations.
Participants	Veteran Israelis ~100 gardeners, most of them over 50 years old.	New immigrants from Ethiopia ~30 gardeners, most of them over 50 years old.
Activity model	Each gardener cultivates his/her own 100 sqm plot. Each gardener has a key and can enter the garden whenever he/she likes. Annual rental payment and other expenses (water, seeds, infrastructure and more). Guidance by an agricultural advisor.	Each gardener is allotted a 10 sqm patch of land, but the community members tend to work together. Only the overseers have the keys, therefore activity hours are limited. No payment. Guidance by a community activist.
Produce	Vegetables, fruit trees, flowers.	Vegetables, herbs (some typical to Ethiopia).
Activity	Working, gathering, social encounters.	Working, gathering, social encounters.

occupations and locations. The following discussion examines the projects' commonalities as reflecting the uniqueness of urban agricultural practices in Israel.

Urban agriculture in Israel

Unlike Europe, the United States and other places around the world, and notwithstanding the short-lived attempt to establish urban farms during the first half of the twentieth century, contemporary urban agriculture in Israel is an emerging practice. The variety of models, many of which have proved to be unstable, demonstrate that urban agriculture in this country is still searching for its unique identity and suitability to the distinctive local conditions summarized below:

- Israel is densely populated, with limited open spaces as compared to common Western per capita standards. The world common standard of 20 sqm of urban open space per capita is relatively rare in Israel. Consequently those who have limited access to these open spaces censure any allocation of urban open space for the benefit of individuals rather than the general public.
- Contrary to many foreign locales, including Western Europe and the USA, vegetables in Israel are relatively inexpensive. Therefore, the economic rationale for growing one's own food is currently negligible, and although commonly stated as a reason for practicing urban agriculture, food security is not as dominant a factor in Israel as elsewhere.
- Scholarly literature positions urban agriculture and community gardening within the framework of contemporary environmental awareness (Wakefield et al., 2007). Therefore, it is not surprising to find Israel's Ministry for the Protection of the Environment and local environmental organizations among the dominant supporters of urban agricultural projects. Still, while the environmental rhetoric is prevalent in both Community Gardens in Israel and Earth's Promise websites, on the ground and in everyday practices it appears that the "green agenda" plays a limited role. Although Garden for the Resident supposedly implements organic principles, such as avoiding the use of chemical pesticides and fertilizers, in practice municipal authorities do not enforce these regulations. In the Ethiopian community gardens environmental issues are promoted by SPNI and other environmental organizations, however these are not a significant aspect of day-to-day gardening practices.

Contemporary urban agriculture in Israel as an emerging environmental paradigm

While current Israeli urban agriculture does not abide by a strict green agenda, one can argue that a holistic environmental paradigm is both at the essence of, and a source of motivation for, the practice.

Eilon Schwartz, former head of the Heschel Center formulated such a holistic paradigm.[4] Based on his discussion of the history of dominant environmental paradigms in Israel, Schwartz suggests that former environmental concepts, such as nature conservation for the sake of nature, and nature conservation for the sake of human wellbeing, are currently replaced by a more holistic environmental paradigm based on three components: place, community and values (Schwartz, 2006).

Place

According to Schwartz, an environmental model based on place represents significant linkage to the Zionist ethos of returning to the nation's ancestral homeland. This claim still prevails today, as evident in the following proclamation by a group of young American Jews who in 2005 established a five-month residential agricultural apprenticeship program in Israel: "Together we are farming Israel, returning from exile to our foundations, grounding ourselves in the midst of a dreamscape. We are planting the letters of our holy scriptures in soil of the land, and we are their roots" (www.jewishfarmschool.org/shorashim.htm.).

Place carries both conceptual and physical meanings. In their seminal essay on the multivalent notion of place – *Makom* – in Jewish Zionist and Israeli thought, Gurevitz and Aran (1991) point to the constant tension between the appeal of Zion as an imaginary locale derived from biblical stories and Jewish legends, on the one hand, and the actual Land of Israel, on the other. Practicing urban agriculture in Rishon LeZion by both veteran Israelis and new immigrants from Ethiopia can be perceived as their attempt to create a place for themselves, a place where they can put down roots in unfamiliar terrain and thereby replace their former imagined concept of place in Israel with a real one.

For senior Ethiopian gardeners, the gardens and agricultural practices invoke mostly memories of their land of origin, Ethiopia. Thus the relationship between the imagined and the real places prove to be more complex: what was an imagined place (the Land of Israel) in the past becomes a concrete place in the present, and what was a concrete place (Ethiopia) becomes an imagined place, one they still yearn for.

Community

Although members of Garden for the Resident cultivate their plots individually, and Ethiopian gardeners cultivate theirs communally, both constitute common communities of origin and shared interests. These communities are separated from nearby neighborhoods, physically by wire fences, and also socially (mainly the Ethiopian community). Contrary to the past, when initiatives were grounded in the Israeli melting pot ethos that aspired to mold all Jews into one nation, these community gardens bolster the cohesiveness of distinct communities. Accordingly, the Garden for the Resident community evolved from a group of strangers, while the Ethiopian Community Garden developed from a group of individual senior Ethiopians who hitherto had little reason to leave their homes. As is common in other immigrant communities, the garden and gardening were found to strengthen this sub-community of seniors who had lost their social status during the immigration process. These two case studies represent a model of exclusive gardening communities, as opposed to more inclusive models that encourage cooperation between veteran Israelis and newcomers.[5]

Values

Though the third component of Schwartz's contemporary environmental paradigm is clearly based on values, he does not identify the specific values involved. Observation of gardening work in Rishon LeZion suggests that it fosters social and environmental values, such as mutual assistance, place connectedness, environmental awareness and others.

Interestingly, these three components: place, community and values, link the practice of urban agriculture to the conceptual framework of what DeLind calls "civic agriculture". According to DeLind, civic agriculture "contains within it the seeds of a more mutualistic and holistic way of being and belonging. It recognizes, as its name implies, commitments that transcend the economic and that privilege citizenship and civic engagement" (DeLind, 2002).

Notes

1   Immigration from Ethiopia started in the 1960s and the process accelerated between the mid-1980s and the early 1990s as more than 20,000 people arrived during two main immigration operations.
2   72 percent of the community's members work, but are poor: 60 percent are under the care of the welfare authorities. According to the Central Bureau of Statistics (2011), the income level of an Ethiopian household is 33 percent lower than that of the general population (www.iaej.co.il/?p=4853#sthash.G5q4Wuw4.dpuf).
3   For example, in the USA various community gardens were established for the benefit of Asian-American communities who settled in America during the late twentieth century, like the Hmong immigrants (Alon-Mozes, 1995). The experience of the Ethiopian immigrants is very similar to the experience of other immigrant groups who found themselves in a foreign country. See for example Saldivar-Tanaka and Krasny, 2004.
4   The *Heschel Center* is dedicated to building a sustainable future for Israeli society – environmentally, socially and economically – through education and reflective activism. Its strategies include leadership development, education, working with government and NGOs, and fostering activism and community-based projects across all sectors of Israeli society.
5   Ethiopian community activists as well as others criticize the state's patriarchal attitude toward the Ethiopian community, which prevents it from integrating into the Israeli society better and more quickly. The role of Ethiopian Community Gardens for this population and their contribution to strengthening its identity or to integrate it with other communities through common gardening practice is an issue for further investigation.

References

Alon-Mozes, T. (1995) Landscape of Immigrants: The Hmong in Merced, Sacramento and Stockton (California), M.L.A. Thesis, University of California, Berkeley, U.S.A.

Alon-Mozes, T. (2007) Women and the Emerging Hebrew Garden in Palestine. *Landscape Research* 32(3): 311–331.

Alon-Mozes, T. (2015) Food for the Body and the Soul, Hebrew-Israeli Foodscapes. In: Imbert, D. (ed.) *Food and the City*. Washington D.C.: Dumbarton Oaks Colloquium on the History of Landscape Architecture, pp. 55–82.

Alon-Mozes, T. and Amdur, L. (2005) *Urban Agriculture in Israel*. Nekudat Chen (in Hebrew).

Alon-Mozes, T. and Amdur L. (2010) Urban Agriculture in Israel: Between Civic Agriculture and Personal Empowerment. In: *Acta Horticulturae*, 2nd International conference of Landscape and urban horticulture, Bologna, Italy, June 2009.

DeLind, L. (2002) Place, work, and civic agriculture: common fields for cultivation. *Agriculture and Human Values* 19: 217–224.

Gurevitz, Z. and Aran G. (1991) Al ha-Makom. *Alpayim* (4): 9–44 (in Hebrew).

Helphand, K. (2006) *Defiant Gardens, Making gardens in wartimes*. San Antonio.

Katz, Y. and Neuman, S. (1996) Women's quest for occupational equality: The case of Jewish female agricultural workers in pre-state Israel. *Rural History* 7(1): 33–52.

Saldivar-Tanaka, L. and Krasny, M. (2004) Culturing community development, neighborhood open space, and civic agriculture: The case of Latino community gardens in New York City. *Agriculture and Human Values* 21(4): 399–412

Schwartz, A. (2006) Changing Paradigms of Environmental Perceptions. Download from: http://heschel.org.il/heshelphp/hachiva1.php?ind=11 (in Hebrew).

Wakefield, S., *et al.* (2007) Growing urban health: Community gardening in South-East Toronto. *Health Promot. Int.* 22(2): 92–101.

14 Population dimensions, land use change and food security in the peri-urban area of Santa-Babadjou, Western Highlands, Cameroon

Roseline Njih Egra Batcha

Introduction

The multiple dynamics of global challenges such as population growth intertwining with land use changes, urbanization, and food security are helping to shape and determine the state of food security challenges. The global population is about 6.8 billion, and it is projected that by 2025, it will reach about 8 billion. Most of the bulk of the additional population is expecting to come from the developing world, which will see its population increase by 61 percent by 2050. World population is growing at an exponential rate and people, be they in urban or rural areas, depend on food for everyday sustenance (UN-ESA, 2009). Population densities of Cameroon stand at 62.6, which are relatively high and are on the increase (World Bank, 2013). Food availability is an important factor affecting human population growth, along with other density dependent limiting factors (Hopfenberg, 2003). The increasing population will therefore have a significant role in influencing and shaping the nature of change and future of land use and food security.

'Hunger has always been an invitation to make a better world, and it remains so' (Roberts, 2008). Throughout the history of human existence, food has held centre stage in the construction of human societies and the evolution of human activity. Food is vital to our understanding of who we are as nations and communities, and the healthy production of food beyond subsistence levels (Budge and Slade, 2009) is a call for attention. This opens up avenues for deeper assessments on linking exponential growth rates in population and food security needs, which may lead to even more transformations of landscapes, land use, land cover, development or shocks within societies. By this, greater concerns over issues of population, land use change, urbanization, climate change, and food security will remain top on the international agenda on global challenges (FAO, 2005).

Population growth influences land use patterns in combination with consumption behaviours and productive activities of the world's people (Wolman, 1993). Current population growth is being accompanied by a change in human habitat. Population growth, migration, family formation, and household fission lead to greater population density and land fragmentation. This growth and density leads to increased consumption of natural resources through the expansion of grazing and fodder, as well as in increased agricultural production (intensification and extensification on marginal lands) (Mortimore *et al.*, 1994). Understanding trends in population size and change are crucial to estimating the future demand for food (Godfray *et al.*, 2010).

Examining and understanding the systemic of present and future human population distribution is critical to understanding the impacts that population growth could have on the rest of the human-environment system in the future. The introduction of satellite image systems presses a new paradigm shift to the forefront of spatial demographic analysis. Satellite data can now be used along with more traditional demographic techniques, providing an even better tool for evaluating population dynamics. It has become one of the strongest tools for analysing and interpreting the complex systems of the earth and the anthropogenic influences that continue to pressure the planet's limited resources. This of course must be taken in the context of inherent data errors and all other problems and assumptions associated with extrapolation from one scale to another (Muhlestein, 2008).

This chapter thus questions how population dynamics affects the sustainability of agricultural land use and also the vulnerability of land users' in being food secure. It seeks to establish a link between land use land cover change and population change dynamics in determining food security.

Material and methods

Study area

The study straddles the north west and western regions, precisely the Santa and Babadjou Sub Divisions, of Cameroon, located between longitudes 10° and 11° and latitude 4° and 6° of the Greenwich meridian (Figure 14.1). It is found in the western highlands of Cameroon, having unique physical entities such as Mount Bamboutos (2740 m), and, it projects into the north west region and covers Mount Lefo in Santa and Mount Oku (3008 m). This area is found within the rapidly expanding cities of Bamenda and Bafoussam whose urban tendencies are fast spreading into the Santa-Babadjou area and peri-urban functions are influencing the land use activities of the area.

The people of Santa-Babadjou are made up of a good number of tribal groups: the Bameléké – Tikars, which are the dominant tribes in the Babadjou area, while the Widikum of the Bantu and Tikars, are dominant in the Santa area. Also, the Fulani and the Mbororo are found in very few numbers comparatively. Both areas of Santa and Babadjou have an inter-mix of the aforementioned tribal groups which can be traced from their similar historic migration origins and through inter-marriage. The Santa people of the north west region, part of former southern Cameroon, had as colonial masters the British, while the Babadjou had the French. Despite different colonial influences, one still finds a similarity in activities of market gardening and in their land use land cover patterns.

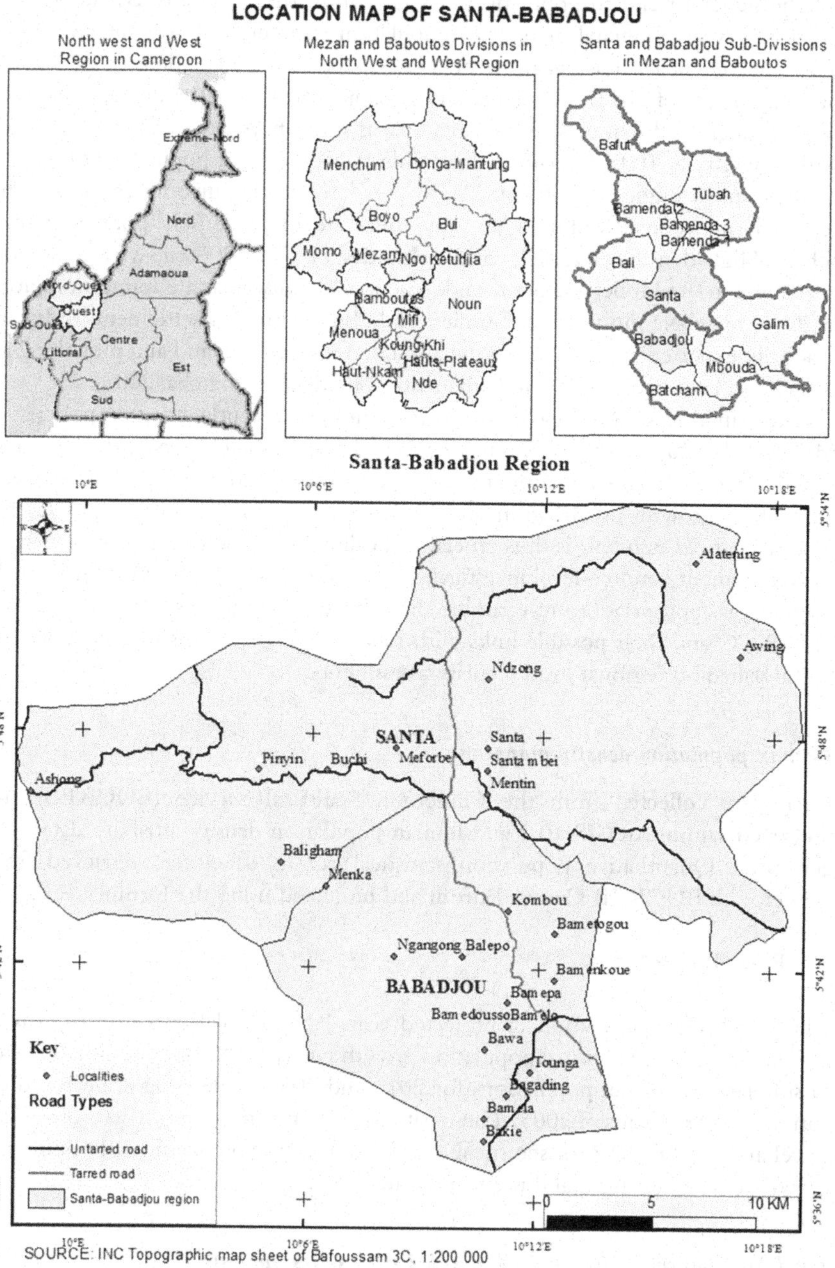

LOCATION MAP OF SANTA-BABADJOU

North west and West Region in Cameroon

Mezan and Baboutos Divisions in North West and West Region

Santa and Babadjou Sub-Divissions in Mezan and Baboutos

Santa-Babadjou Region

SOURCE: INC Topographic map sheet of Bafoussam 3C, 1:200 000

Figure 14.1 Location of Santa-Babdjou

The general trend in population change within the Santa and Babadjou subdivisions reveals a general increase in population numbers and densities over time. This area has a total population of 142,767 persons and a density of 210.84 persons/km^2 as of 2013. The Santa and Babadjou areas do have different population numbers and densities. The Santa area is more populated at 76,750 persons with a density at 160.67 persons/km^2, while Babadjou has a population of 66,017 persons and a higher density of 221.36 persons/km^2. Population density of the region is unevenly distributed, as some areas are more densely populated than others. High-density areas fall within altitudes of 1500–2000m, where in most settlement-related functions are found. The low-density areas are found in regions at 2000+m, where most grazier families reside. Such dispersed settlements today are gradually experiencing growing population densities on the land and provoking an increase in land occupancy with urbanisation tendencies (Batcha, 2008).

The Santa-Babadjou areas have been earmarked recently by state policies of MINADER (2013) as regions with high need for regional analyses of food security challenges. An investigation, recording and understanding of the link between population growth, urbanisation and land use land cover change (LULCC) for food security assessments is thus crucial and calls for the need for regional analyses to complement country-level investigations to reduce posterior food security challenges. It is appropriate to re-examine the issue and dimensions of the root causes of LULCC and their possible links with food security, which is missing from the Santa-Babadjou regional food security assessments.

Human population density mapping

Data were collected from the Cameroon Statistical Service (BUCREB) and processed within ArcGIS 10.2 and human population density attribute data were generated. Quantitative population statistical census data were retrieved from Cameroon's BUCREB Census Bureau and projected using the formula:

$$P = P_o (1+A)^y$$

where P = total population of projected year; P_o = population of initial year of simulation; A = % of annual population growth rate; and ^{y} = exact number of years of simulations. To get populations for 2013 and 2050, projections were based on national census results of 2005. These data were then exported to Microsoft Office Excel and ArcGIS 10.2 for statistical analysis and for the preparation of population growth, density change/land use time series graphs.

Land use land cover change detection and time series analysis

Land use land cover classification was done using landsat satellite images of 1973, 1988, 2001, and 2013. They were processed in ArcGIS 10.2 using the ISO Unsupervised classification function to extract the land use land cover classes. Five main classes were identified: forest cover, mixed agricultural land use, built-up, bare land surfaces, and water. Then change detection was run comparing the different land use classification data sets. Attribute data of the classification data sets were used for correlation with demographic data of the Santa Babadjou area.

Food security assessments

Food density mapping

The food density maps combined current and projected per capita food availability (calories/person/km²/day) and growth in food consumption expressed in daily calories available per person per day for the region with current projected population densities and population growth data.

Per capita food consumption is thus:

$$\frac{\text{Total food consumption (calories/person/day)}}{\text{Total population (persons)}}$$

The food availability records from MINADER were incomplete, and using it for estimation purposes proved difficult, so analysis was based on data from FAOSTAT and NRCE (Natural Resources Consulting Engineers) (FAOSTAT GAUL, 2008) comprehensive reports to extract the food density status and spatial representation for the Santa-Babadjou region of Cameroon.

The population growth data and densities were from BUCREP 2005 statistics and projections were calculated for 2050. The food availability index data were retrieved from FAOSTAT 2008 reports and reprocessed within ArcGIS 10.2. Spatial food density maps and attribute data were generated.

Agriculture land use per capita index

Agricultural area by population numbers of a given time indicates an area's available land resources for producing its own food:

$$\text{Agriculture land use per capita index} = \frac{\text{Agricultural area}}{\text{Population number}}$$
(Available agricultural resources for producing food)

Land use land cover classification maps were used to get an estimation of the area for agricultural land use for the different time periods. Projected total agricultural land area for 2050 was based on projections and assumptions of the possible state of land use at time (*t*).

Food frequency assessments and household consumption rankings

In understanding household food security perspectives, household demographics and socio-economic characteristic were identified for the different household members through a household survey. A total of 600 households were visited for the general household survey and 300 households represented for in-depth interviews. Food frequency assessments using the household consumption security rankings method involved the collection of minimum amounts of food consumption data from the household members. Though this method is limited in its level of precision, we found it cost effective and it helped us to detect the consumption differences between the households.

Results and discussion

This study assessed the population dimensions of land use change and food security in the peri-urban area of Santa-Babadjou in the western highlands of Cameroon. Findings reveal that from 1973 to 2050 increased urbanization tendencies, population numbers and densities, changing land use land cover patterns, and competing land uses over agricultural land use are greatly altering the human environment and posing a challenge to the food security status of the Santa-Babadjou communities.

Population change and land use

The general trend in population change within the Santa and Babadjou sub-divisions reveals a general increase in population numbers and densities over time. In 1973, the population stood at 49,721 persons and a density of 74.53; in 1988, the population rose to 91,415 persons and a density of 135.00. This trend continued in 2001, with a population increase of 92,112 persons and a population density of 136.03, to 2005, having a population of 103,320 persons and a density of 152.58. From 1973 to 2005, the population experienced a doubling and a tripling is projected to happen from 2005 to 2050, with 2050 projected to have a population of about 307,533.

Linking population change and land use change, population change correlates with changing land use patterns in the Santa-Babadjou region of Cameroon. The extent of competitive occupation of bare land and open grazing fields over time shows the extent of growth in human population numbers. This progressive occupation of bare land and open grazing fields by either agricultural activities or by the built-up environment is very evident in this region. Also, it was noticed that the greater the population, the greater is the expansion of urban development from dispersed to dense settlements, or from the traditional built environment to a more modern structured environment. Statistics from the net changes in mixed agricultural land occupation, built-up land occupation (Table 14.1) reflects the extent of people's increased use of the land's resources. Population density is increasing and more forest and bare land are increasingly used for settlement purposes.

Agricultural land use is challenged to meet the demands of the population that is urbanizing and growing in number. Agriculture has been and is still the mainstay and a very important activity in the Santa-Babadjou region. This activity is dominant in the plains and with little activity in the uplands, it is also a predominant activity around settlement sites and along the

Table 14.1 Change in population and land occupation: 1973–2013

Attributes	Population 1973	1988	2001	2013
Santa-Babadjou	49721	91415	92112	142767
Surface area (Km²/Ha)	677.13	677.13	677.13	677.13
Densities	74.53	135.00	136.03	210.84
Forest land	31674	504236	97955	281331
Bare land	84715	173073	177240	165251
Mixed agricultural land	57932	450598	697301	727311
Built-up land	10249	259989	300472	222950
Water surfaces	2341	44443	121868	105899

Source: based on BUCREP (2005).

Matezem, Lessegue, and Tsoumbang rivers. Transformation of agricultural land to residential land due to increase in population and also as a result of the expansion of nearby towns of Bamenda and Bafoussam is seriously challenging the sustainability of agricultural land and the food security needs of the population of this peri-urban area.

Food security assessments

Agricultural land use per capita index analysis

Agricultural land use per capita index analysis gives a measure of the amount of pressure the population has on available resources. It differs from area to area and has been changing over time in this region. The Santa area has less pressure on its resources as compared to the Babadjou area. This information correlated with the population density data of these areas reflecting the fact that Babadjou, with more dense population, has a higher pressure on its available agricultural resources compared to the Santa area.

Comparing Santa-Babadjou region with an agriculture area per capita index in 2013 of 5.09 persons/ha to the national average which stands at <0.5 persons/ha, we find that of Babadjou to be relatively high. Though much land area relatively is used by agriculture, other land uses reveal that high competition over the use of land poses a future problem for the availability of land for agricultural purposes. The available area per capita index rose rapidly from the 1973 (1.16 persons/ha/km²) to 7.1 persons/ha/km² in 1988 and to 7.57 persons/ha/km² in 2001, and took a downward trend of 5.09 persons/ha/km² as of 2013. If we assume an increase in population with little increases in agricultural land area, a linear forecasting line reveals that the available area per capita will drop even more into the future, demanding that we think of new agricultural strategies.

For 2050, we examined two scenarios: first, we assumed that the agricultural land area does not change as of the year 2013, meaning no increase or decrease in total agricultural land area. Second, we assumed that the bare land surfaces were

Table 14.2 Available area per capita indexes (1973–2013) and projected agricultural resources for 2050

Attributes of available area per capita index	Year 1973	1988	2001	2013
Agricultural land resources (ha/km²)	57932	650598	697301	727311
Population count (persons)	49721	91415	92112	142767
Available area per capita indexes	1.13	7.1	7.1	5.09
Area per capita index attributes **Available agricultural land resources (Ha/Km²)**			**Area per capita index: 2050**	
Scenario 1				
Agricultural land area as of 2013 LULC classification	727311		2.36	
Scenario 2				
Bare land area of 2013 and available agricultural land area (2013)	65251+727311= 892562		2.9	
Population count of 2050 (persons)	307533			

Source: based on BUCREP (2005).

converted to agricultural land area due to increased need for more farming to accommodate the needs of the growing population. Also, the people maintained both traditional and agricultural related economic activities as their mainstay. From the evaluations based on these assumptions, only a slight increase in available area per capita index is realised and from the first assumption it stands at 2.36, while from the second assumption the index is 2.9.

From the trends in available land per capita index, we observe a steady rise from 1973, 1988 to 2001 (1.13 to 7.1 to 7.57), and then to 5.09 in 2013 and to 2.36 in 2050. This implies a decrease in land per capita index, which reflects an adverse effect on crop productivity and availability, which are key determinants of food security measures.

If we factor in other land use change variables, such as climate variability, land degradation, competing land uses, which may occur on the land given its finite state, then available area per capita index, which is currently low in the Santa-Babadjou region, with high population densities and relatively low available agricultural resources, makes the region prone and vulnerable to food insecurity if adaptive measures are not put in place now and in the long term.

Food density mapping analysis

The food density spatial data for the Santa-Babadjou region for the periods 2005 and 2050 (Figure 14.2) combines the current and projected per capita food demand (calories/person/day), with the current and projected population densities (persons/km²). Food density mapping reflects the current and projected data on food consumption density (calories/person/km²/day). Food density mapping and

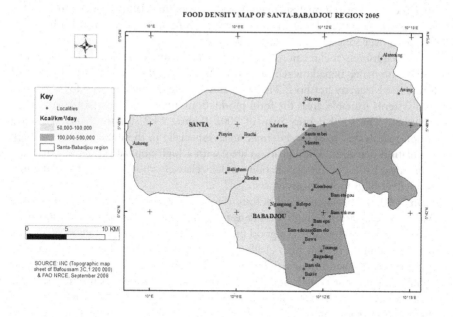

FOOD DENSITY MAP OF SANTA-BABADJOU REGION 2005

FOOD DENSITY MAP OF SANTA-BABADJOU REGION 2050

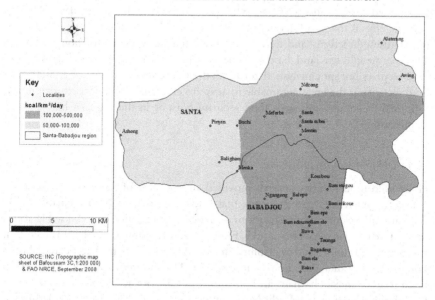

Figure 14.2 Food density maps of Santa-Babadjou: 2005 and 2050

food security hotspot analysis made use of agricultural land use per capita index analysis which revealed that this index differs from area to area and also has been changing over time in the Santa-Babadjou region.

In establishing the relationship between population and land use tendencies within the Santa-Babadjou region and food security, the study revealed potential future food security hotspots. These are areas where the pressures on food production systems (suitable areas for food production) are and will be particularly high in future. The food density map of 2005 (Figure 14.2) for the Santa-Babadjou region reveals that high food density areas, especially in the Babadjou, reflect high population densities. The Babadjou area, with a high population density than the Santa area, has a low food availability index relatively, though some parts of Santa also reflect high food density states.

In 2050, the projected food density (Figure 14.2) situation reveals an increased spatial representation of high food density, spanning all of Babadjou and most parts of Santa area. Food density is a good indicator of the food security status of an area: when it is high it reflects high challenges to food security or the presence of food insecurity, and when low, it reflects low food security challenges. The main driver behind this forecast is high population growth and increase urbanization tendencies. Pressure is on the available agricultural land resources and food systems and if these are not operating well, this may cause this region to be a potential food security hotspot in Cameroon.

Food density maps are somewhat biased due to the assumptions and projected data used, which does not reveal the actual situation. The food availability data were based on national averages, which do not reveal the actual data. What is advantageous is that food density mapping gives us an indication of the food security statuses, directions, and trends, which are necessary inputs for policy making. The way people use land, based on the agricultural land index, food density, and food security hot spot analysis, determines the responses and concerns about future food security. If the areas are unable to adapt to changing conditions, the region in time may become a potential food security hotspot in the western highlands of Cameroon.

Household food security perspectives

Household food security analysis provided an overview of the various indicators, conceptual refernces, and processes that lead to household food security (HFS). Assessing HFS demands investigations into food supply, availability, and stable access to food, which are very critical to HFS status. Information at this level is based on the factors that limit food availability and the different options that are available to households in Santa-Babadjou to access food. The indicators used here depended upon the financial, human, and institutional support resources available. The household characteristics, both demographic and socio-economic, provided much needed baseline information on food security status: 75 percent of the households were involved in agricultural activities, with crop cultivation (60 percent) and livestock herding (15 percent) as the main occupations. Non-agricultural activities employed about 25 percent of the population, with traders at 10 percent and public sector workers at 15 percent.

The main source of food for the households in Santa-Babadjou is the produce from their farms. More than 65 percent of the population depends on crop production and the rearing of animals to secure livelihoods and ensuring food security. Domestic food production, as compared to food importation, plays an important role in food security in the Santa-Babadjou region. Most of the households rely on the domestic production of food and also rely on the market for purchase of food supplies. The markets serve as a supply bank for foodstuffs in times of seasonal scarcity or to compliment those food items that are not produced on the farm. Also, these households rely or turn to the markets when they do not have sufficient quantities of food to meet demand during the off harvest periods.

Most farmers in Santa-Babadjou sell most of their produce to exporters, especially grains and vegetables. They do so to be able to cover production expenses and other livelihood needs such as clothing, celebrations, school fees, and household equipment. Another important reason that causes farmers to sell most of their produce is the high influx of consumers from neighboring countries such as Nigeria and Gabon. These exporters buy directly from the farms and pay relatively better than the locals. This is a critical factor that determines the status of household food availability.

Food frequency assessments

Also at the household level, food frequency assessments, which involved the collection of food consumption data, revealed that 71.67 percent of households fear running out of food, especially during off-harvest seasons. The average number of

Table 14.3 Sources of household food supplies and household food frequency assessments

Sources of household food	Number of respondents		Unit (%)
Farm production only	90		15
Farm production and market purchase	450		75
Market purchase only	42		7
Others	18		3
Total	600		100
Household consumption security rankings	Frequency (number)	Units (%, mean)	Rankings
Number of meals eaten per day	600	2	Low
Number of meals per week which includes meat or fish	600	3	Low
Number of ingredients in the sauce or food per meal	510	4	Average
Number of times a week a nutrient poor meal was prepared as a main meal	600	4	Low
Fear that food will run out of supply	430	71.67%	Low
Ration always sufficient for household members	320	53.33%	Low

meals eaten per day stands at an average of two, which is less than the normal three meals per day. Of these households, only 53.33 percent reported that their daily rations had always been enough for their household members. Also, in terms of number of meals with either meat or fish, most of the households attested to eating meat or fish about three times a week, especially on market days. From these statistics, a good indication of prevalence of food insecurity is noted in the region.

Examining household perception of food security, the household consumption security rankings revealed that:

- Most of households (71.67 percent) are experiencing low food consumption security. This standard is most prevalent during those periods with seasonal food shortages.
- Also 53.33 percent of households have low consumption security because they do not always have sufficient food for all their household members.
- In light of number of meals eaten a day and those with low nutritional standards, households in the Santa-Babadjou region reflect low to average food consumption security.

The results also show that even when food maybe available at all times for household members in enough quantities, there is no guarantee that the diet may meet the nutritional standards needed to be food secured. From the above statistics and findings related to household consumption security rankings, we found a significant positive determinant relationship between food availability and food security status of households in the Santa-Babadjou region.

Food frequency assessments using the household consumption security rankings method are limited in their level of precision, but the method was cost effective and it helped us to detect the consumption differences between the households. These findings thus provide the evidence that food security continues to be a social problem for the Santa-Babadjou population and for the development of the region. This therefore demands that policy makers to pay close attention to food production and agricultural land availability issues at this regional level to meet the needs of this growing peri-urban area.

Conclusion

The multiple dynamics of population growth intertwining with land use changes, urbanization, and food productivity are helping to shape and determine the state of food security challenges in the Santa-Babadjou peri-urban region of Cameroon. Food security hotspot analysis, high population densities and growth, as well as low food per capita availability, reveals that agricultural land availability is compromised by other competing land uses, most especially by urbanization tendencies. Also, food security is under threat as food availability in this region, which is dependent on the productivity of agricultural lands, is not consistent all year round and is highly impacted by changing population dynamics and by changing land use land cover.

It is helpful to conclude that population growth and change lead to land use changes and land fragmentation. These multiple land uses are challenging and limiting the availability of agricultural land and posing a threat to food security both currently and in the future. Since the population is predicted to rise, agricultural land use remains an important land use activity in the Santa-Babadjou peri-urban region. Policy makers are thus called upon to ensure sustainable land use planning to adapt to the changing conditions of urbanization and land use, and to prevent the whole region from becoming a potential food security hotspot.

References

Batcha, R.N.E. (2008) Land use dynamics in Santa-Babadjou area, Cameroon. DEA Dissertation, University of Yaoundé 1, Cameroon.

Budge T. and Slade, C. (2009) *Integrating Land Use Planning and Community Food Security: A New Agenda for Government to Deliver on Sustainability, Economic Growth and Social Justice.* Community Planning and Development Program, LaTrobe University

FAO (2005) Farming in urban areas can boost food security. FAO Newsroom. www.fao.org/newsroom/en/news/2005/102877/index.html).

FAOSTAT GAUL (2008) *FAO NRCE Population Dataset,* UN population Division World Agriculture: towards 2030/2050, Projection: Geographic projection, WGS 8 4.

Godfray I., R. Crute, H. Charles, J. Lawrence Haddad, D. Lawrence, J.F. Muir, N. Nisbett, J. Pretty, S. Robinson, C. Toulmin and R. Whiteley (2010) *The Future of the Global Food System.* Royal Society, London.

Hopfenberg R. (2003) Human carrying capacity is determined by food availability, *Population and Environment* 25(2): 109–117.

MINADER (2013) *End of Year Report* (Unpublished manuscript).

Mortimore, M., M. Tiffen and F. Gichuki (1994) *More People, Less Erosion: Environmental recovery in Kenya.* John Wiley and Sons, Chichester.

Muhlestein K.N. (2008) *Land Use Land Cover Change Analysis of Maverick County Texas along the US Mexico Border.* EES 5053 Remote Sensing, University of Texas at San Antonio Environmental Science and Engineering PhD Program.

Roberts, P. (2008) *The End of Food: The Coming Crisis in the World Food Industry.* Bloomsbury Publishing, London.

UN-ESA (2009) *World urbanization prospects: The 2009 revision.* UN-ESA. www.ctc-health.org.cn/file/2011061610.pdf.

Wolman, M.G. (1993) Population, land use and environment: A long history. In: Jolly, C.L. and B.B. Torrey (eds) *Population and Land Use in Developing Countries.* National Academy Press, Washington, DC.

World Bank (2013) World development indicators 2013. World Data Bank databank. http//worldbank.org/data/download/WDI-2013-ebook.pdf.

15 UK farming entrepreneurship for food security in an uncertain future

Howard Lee

Introduction

European agriculture and its farmers, production horticulture and its growers are currently facing many pressures: climate change, reduced resource availability, responsibilities to protect the environment, volatile prices and ever-moving political agendas. Whilst farms provide both goods and services, an immediate issue for the future is guaranteeing food supplies in an ever more uncertain world, thus, the ability of Europe to feed itself is becoming ever more important (Candel, *et al.*, 2014). In the UK self-sufficiency has already been reviewed (Lee, 2015) and shown to vary for fresh produce, from 8–50 per cent for fruit and vegetables, respectively. Whilst some commodities such as wheat are abundantly produced in the UK, overall self-sufficiency is lacking, even though UK farmers currently manage relatively larger scale and more intensive farming systems than for most of Europe (Kempen, *et al.*, 2011). Whilst economies of scale have been established for larger farms (Gadanakis, *et al.*, 2015a) (Keizer & Emvalomatis, 2014) there is a continuing interest in smaller, localized food production (Kirwan & Maye, 2013). This chapter attempts to evaluate the potential future role of small-scale entrepreneurship for a more food secure UK as we face an ever more uncertain future.

UK food productivity and entrepreneurship

In recent years to enhance UK farm production has been a key government ambition and the Department of Environment, Food and Rural Affairs (Defra) and other key advisers have suggested a policy of 'sustainable intensification' (SI). Gadanakis *et al.* (2015b, p. 288) define SI as to: "simultaneously raise yields, increase input use efficiency and reduce the negative environmental impacts of farming systems to secure future food production and to sustainably use the limited resources for agriculture". These authors have attempted to assess British farms in terms of an 'index of Eco-Efficiency,' but admit that:

> the design of agricultural policy which aims to achieve sustainable intensification is a difficult and complex procedure because a) it requires encouraging farmers to change their attitudes and behaviors such as adopting new

management practices; and b) there are unpredictable external factors such as weather, disease and input costs variability.

(Gadanakis *et al.*, 2015b, p. 297)

In a USA study, SI as a 'silver bullet' solution has also been treated cautiously, with the exhaustion of finite supplies of phosphorus seen as a key-limiting factor (Petersen & Snapp, 2015).[1] These authors interviewed agricultural experts about SI, who emphasized a lack of understanding about agro-ecosystem function and the need for more research to help enhance agro-ecological productivity (Petersen & Snapp, 2015). This is part of an understanding of agro-ecosystems that comprise an interaction of environmental, social, economic and political factors (Dalgaard, *et al.*, 2003).

The social (human) dimension of agro-ecosystem function is of interest for this review: *i.e.* how can farmer behavior help innovate and develop their farm businesses to enhance yields? Business innovation is now redefined as entrepreneurship – a term which has recently been reviewed by Lee (2015) as "the process of uncovering or developing an opportunity to create value through innovation" (Henderson, 2002, in Lee, 2016).

Entrepreneurship for food production in the UK has interested policy makers in the past decade, with the encouragement of many new farming enterprises such as rural office space, private storage, fish farming, wind turbines, solar arrays, eco-tourism etc. But innovation for on-farm food production has been less widespread. A well-established environmental management kite mark is operated by LEAF[2] and other on-farm schemes have been reported such as High Nature Value Farming (Strohbach, *et al.*, 2015) and Care Farming (Leck, *et al.*, 2014). However Norman (2015, p. 1496) indicates that: "Increasing productivity in an ecologically sustainable manner is going to be a complex process...[and] further farmer empowerment and ownership of the initiatives identified and implemented will undoubtedly become critically important in determining whether the productivity and ecological sustainability goals will be met". On-farm complexity is further emphasized by Rodriquez and Sadras (2011).

A key factor is seen as decision making by farmers:

> Irrespective of their level of endowment, farmers use incomplete or imperfect knowledge to make technical (e.g., agronomic, energy inputs, irrigation scheduling, etc.) and financial (e.g., marketing, loans, and off-farm investment) decisions, in a context of risk and uncertainty associated with climate variability, market volatility and political and global change.
>
> (Rodriquez and Sadras, 2011, p. 138).

So, can SI help us deal with this level of complexity and help farmers grow more food in a climate of uncertainty?

Planning to grow food in a climate of risk and uncertainty

Barnes and Thomson (2014) attempt to rationalize the implications of SI for social risk, shown from their paper (Figure 15.1) as a risk-return framework, i.e. exploring the relationship between the return (yield increases) and (social) risks.

Based upon Figure 15.1 Barnes and Thomson (2014, p. 214) state that, for *any* farm technology:

> The only option for improving yield is to intensify production, that is move from point A to point C. Clearly, this increases the social risk, for example more resources are needed to produce more yield which, even on the efficiency frontier leads to an increase in the potential for damage.

The authors suggest that: "the only option is to see sustainable intensification as a new technology. This is represented by the dotted line where a farm can increase yield by shifting up to point E. Here yield has increased with no increase in social risk". However, they suggest that moving along this new frontier (E to G) will lead to increased social risk. They comment that SI "must be considered a new technology which reconfigures the relationship between inputs and outputs and effectively raises the frontier continuously upward" (p. 214).

But, Barnes and Thomson do not clarify how effective SI would be at coping with external constraints and negative impacts. It is suggested in this review that we need production systems that can function with potentially reduced inputs and recover from negative external shocks to still give useful yields. In Britain there

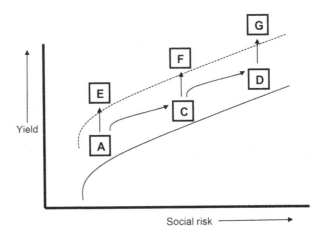

Figure 15.1 Risk-return framework applied to sustainable intensification

Source: Redrawn from Barns and Thomson, 2014, p. 214.

have been some excellent studies to balance optimal yields with reduced inputs. The (then) Ministry of Agriculture, Fisheries and Food (MAFF) (now Defra) funded the Boxworth project (1981–1991): the first long-term farm-scale project to investigate more environmentally friendly approaches to crop production in the UK. MAFF subsequently commissioned the TALISMAN and SCARAB research projects (1990–1998) as follow-on studies to investigate the Boxworth results in more detail (Young, *et al.*, 2001). It is strongly recommended that these studies be re-visited as part of a better understanding of how to reduce inputs for UK farming without suffering yield penalties.

In addition to reduced inputs, the *resilience* of farm systems needs consideration. This is reviewed by Barling, *et al.* (2008, p. 3):

> Recent debate has explored the 'resilience' of the food supply — its ability to prevent, withstand and recover from serious shocks. This report suggests that contingency planning[3] is inadequate to the challenges now facing the food system, presented here as the New Fundamentals. These are the framing realities that policies concerning food supply must in future address. The issues include climate change; water; biodiversity and eco-systems support; energy and non-renewable fossil fuels; population growth; land use; soil; labor; and dietary change and public health.

A recent paper (Roege, *et al.*, 2014) reviews resilience in terms of energy provision. They tend to agree with Barling *et al.* that "the current ad hoc approach to system hardening, which typically seeks to address past failure scenarios, does not necessarily assure protection from unexpected future scenarios". They refer to the USA Department of Homeland Security, which states that:"Having accurate information and analysis about risk is essential to achieving resilience. Resilient infrastructure assets, systems, and networks must also be robust, agile, and adaptable", and a National Academy of Sciences report, which identifies: "four basic resilience components: **plan/prepare, absorb, recover**, and **adapt** to anticipated and unanticipated conditions".

Another important paper (Aven & Krohn, 2014) attempts to further address this issue, looking for: "...an effective instrument for managing risks, the unforeseen and potential surprises" (p. 2). They settle upon the '(collective) mindfulness concept'. This is included in their development of an integrated risk perspective (Figure 15.2).

However, food supply in the UK is not based merely on domestic production but also international trade, which is additionally vulnerable to uncertainty and interruptions. A Defra report (2010) *UK Food Security Assessment: Detailed Analysis* confirms that: "For an open, trading economy like the UK, a secure global food supply ultimately underpins long term availability and prices in the UK, although it also generates risks". These are listed as:

- Demographic change;
- Climate change, leading to harvest shortfalls, pests and diseases;

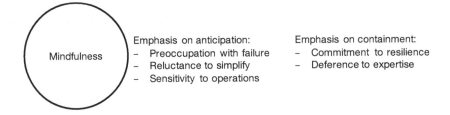

Figure 15.2 Main building blocks for the integrated risk perspective and the new way of thinking about risk, emphasizing the process from fundamental concepts and principles, to an improved understanding of risk and adequate actions

Source: Adapted from Aven & Krohn, 2014, p. 7.

- Third country protectionism and economic nationalism;
- Sustainable productivity growth;
- Ensuring adequate investment and applied research.

(Defra, 2010, p. 12)

Entrepreneurship and food security

Can the encouragement of small-scale entrepreneurship for food production contribute to greater food security against uncertainty? In a series of case studies for small entrepreneur businesses (Lee, 2105), results indicated that training, supportive networks and financial assistance were the most critical factors for success. It was also suggested that many existing projects needed more concerted assistance *via* free or low cost training and attractive finance schemes. Finally, both for new and existing businesses, it was concluded there was a need for better-coordinated support networks, to allow experiences to be shared and successes to be replicated. Whilst UK farmers are generally very well educated compared with the rest of the European Union (EU) there is an ongoing need for further up-skilling as noted above by Rodriquez and Sadras (2011), especially as we face an ever changing and uncertain future.

It is suggested here that a new approach to UK farming is required: the development of 'silver bullet' technology for developing regions of the world has already been eschewed as inappropriate to local conditions (e.g. Tittonell and Giller, 2013) for African smallholders. In the UK such technological solutions to farming have increased yields enormously in the past 100 years (Young *et al.*, 2001) and they still have a vital role to play in short-term food provision but are fragile to sudden external shocks and changes as discussed above. So, planning for future productivity of UK agriculture is a different issue: we need to consider how we can develop farming systems which are *more* than resilient, i.e. which can embrace future uncertainty and yet prosper, to give enhanced yields.

This ambition might be considered unrealistic but there are some hopeful ideas from Taleb (2013) who has developed the concept of *anti-fragility*. This can be summarized from a recent interview in *New Scientist* magazine. Taleb is quoted as follows with bullet points to aid clarity:

- When you ask people what is the opposite of fragile, they mostly answer something that is resilient or unbreakable – an unbreakable package would be robust;
- However, the opposite of fragile is something that actually gains from disorder ... Nature builds things that are anti-fragile. In the case of evolution, nature uses disorder to grow stronger;
- If anti-fragility is the property of all these natural complex systems that have survived, then depriving them of volatility, randomness and stressors will harm them;
- If you want to move away from fragility, you must avoid centralization ... I want the entrepreneur to be respected ...[we should] foster aggressive risk-takers who are willing to fail;
- Trial and error is an anti-fragile activity.

(adapted from Gedded, 2012, p. 30)

These ideas are revolutionary: can we apply anti-fragile thinking to planning future UK farm systems? Archuleta (2014) suggests that, for the USA:

The current fragile [farming] system uses separable, directed research when it comes to science; however the anti-fragile system uses a holistic approach, with tinkering instead of wholesale actions ... When mistakes are made in fragile operations, they are large, irreversible mistakes, which can run up a lot of debt ... [Alternatively, farmers can] embrace anti-fragile operations [by] making little mistakes through experimentation and a little tinkering. These mistakes are often reversible and easily overcome.

Can we learn from these ideas to successfully plan for future food security in the UK? Taleb is especially interested in entrepreneurship, which he sees as a vital factor for anti-fragility (Geddes, 2012). The importance of entrepreneurship has already been discussed above and separately as a potential contributor to urban food security (see Lee, 2012, in the Proceedings of an earlier AESOP Food Planning Conference).

Entrepreneurship to achieve food security?

It can therefore be postulated that longer term food security in the UK might be best achieved by means of an anti-fragile approach which seeks for complexity and diversity: numerous, small projects with a diverse array of crops and livestock being managed using multiple production systems and with a strong entrepreneurial approach. This would require support:

- Active collaboration with existing schemes, projects and organizations – this should be open-handed and acknowledge existing skills and human capital;

- The provision of low-cost or free training workshops, courses and mentoring to help up-skill a diverse array of local and incoming entrepreneurs (Lee, 2012);
- Access to available and attractive finance initiatives to help new entrepreneurial projects prosper.

The first criticism of this approach is that it may be relatively less productive than more coordinated and unified policies: i.e. if food security is so important, why move away from conventional production systems which are based on the latest scientific research to maximize productivity? Could more diverse, local projects become equally productive and potentially more reliable in the face of uncertainty? The suggestion here is not for revolutionary change in British agriculture but instead some funded case studies to critically examine and demonstrate the potential strengths and weaknesses of more heterogeneous localized production systems.

Conclusions

The current UK farming policy of sustainable intensification may not hold all the answers to achieving future food security in the UK in a context of greater risk and uncertainty. Existing data bases for prior studies on reduced input production systems need to be re-visited to help in future planning, and new demonstration case studies are also required to critically examine the potential value of smaller-scale projects managing a more diverse array of crops and livestock, employing multiple production systems and with a stronger emphasis on entrepreneurship.

Notes

1 Phosphorus on farms and global food security is excellently reviewed by Cordell, *et al.* (2009).
2 Linking Environment and Farming: www.leafuk.org/leaf/home.eb
3 Contingency planning: an assessment of the likely effects of a series of possible scenarios on the system being studied – so for farm systems, scenarios might include serious weather events, sudden resource shortages etc.

References

Archuleta, R. (2014) *Archuleta discusses importance of antifragile farming systems.* [Online] Available at: www.theprairiestar.com/news/crop/archuleta-discusses-importance-of-antifragile-farming-systems/article_d8f1a924-9277-11e3-ad9e-0019bb2963f4.html [Accessed 21 May 2014].
Aven, T. and Krohn, B.S. (2014) A new perspective on how to understand, assess and manage risk and the unforseen. *Reliability Engineering and System Safety,* Volume 121, pp. 1–10.
Barling, D., Sharpe, R. and Lang, T. (2008) *Rethinking Britain's Food Security,* Bristol: The Soil Association.
Barnes, A.P. and Thomson, S.G. (2014) Measuring progress towards sustainable intensification: How far can secondary go/. *Ecological Indicators,* Volume 36, pp. 213–220.
Bedore, M. (2014) The convening power of food as growth machine politics: A study of

food policymaking and partnership formation in Baltimore. *Urban Studies,* Volume 20(10), pp. 1–17.

Blay-Palmer, A., Knezevic, I. and Spring, A. (2014) Seeking common ground for food system transformation. *Dialogues in Human Geography,* Volume 4(2), pp. 185–189.

Campbell, M.C. (2004) Building a Common Table. The Role for Planning in Community Food Systems. *Journal of Planning Education and Research,* Volume 23, pp. 341–355.

Candel, J.J., Breeman, G.E., Stiller, S.J. and Termeer, C.J. (2014) Disentangling the consensus frame of food security: The case of the EU Common Agricultural Policy reform debate. *Food Policy,* Volume 44, pp. 47–58.

Cardon, M.S., Gregiore, D.A. Stevens, C.E. and Patel, P.C. (2013) Measuring entrepreneurial passion: Conceptual foundations and scale validation. *Journal of Business Venturing,* Volume 28, pp. 373–396.

Cleveland, D.A. *et al.* (2014) Local food hubs for alternative food systems: A case study from Santa Barbara County, California. *Journal of Rural Studies,* Volume 35, pp. 26–36.

Cordell, D., Drangert, J.-O. and White, S. (2009) The story of phosphorus: Global food security and food for thought.. *Global Environmental Change,* Volume 19, pp. 295–305.

Curtis, F. (2003) Eco-localism and sustainability. *Ecological Economics,* Volume 46, pp. 83–102.

Dalgaard, T., Hutchings, N.J. and Porter, J.R. (2003) Agroecology, scaling and interdisciplinarity. *Agriculture, Ecosystems and Environment,* Volume 100, pp. 39–51.

Defra (2010) *UK Food Security Assessment,* London: Defra.

Dixon, S.E. and Clifford, A. (2007) Ecopreneurship – a new approach to managing the triple bottom line. *Journal of Organizational Change Management,* Volume 20(3), pp. 326–345.

Draghici, A. and Albulescu, C.T. (2014) Does the entrepreneurial activity enhance the national innovative capacity?. *Procedia – Social and Behavioral Sciences,* Volume 124, pp. 388–396.

Gadanakis, Y., Bennett, R., Park, J. and Areal, F.J. (2015a) Evaluating the Sustainable Intensification of arable farms. *Journal of Environmental Management,* Volume 150, pp. 288–298.

Gadanakis, Y., Bennett, R., Park, J. and Areal, F.J. (2015b) Improving productivity and water use efficiency: A case study of farms in England. *Agricultural Water Management,* Volume 160, pp. 22–32.

Geddes, L. (2012) Lift weights, avoid debt, drink the water.. *New Scientist,* 17 November, pp. 30–31.

Glavic, P. and Lukman, R. (2007) Review of sustainability terms and their definitions. *Journal of Cleaner Production,* Volume 15, pp. 1875–1885.

Goerner, S.J., Lietaer, B. and Ulanowicz, R.E. (2009) Quantifying economic sustainability: Implications for free-enterprise theory, policy & practice. *Ecological Economics,* Volume 69, pp. 76–81.

Henderson, J. (2002) Building the Rural Economy with High-Growth Entrepreneurs. *Economic Review,* Volume third quarter, pp. 45–70.

Jarratt, D.J. (1996) A comparison of two alternative interviewing techniques used within an integrated research design: a case study in outshopping using semistructured and nondirected interviewing techniques. *Marketing Intelligence & Planning,* Volume 14(6), pp. 6–15.

Jones, M.V., Coviello, N. and Tang, Y.K. (2011) International Entrepreneurship research (1989–2009): A domain ontology and thematic analysis. *Journal of Business Venturing,* Volume 26(6), pp. 632–659.

Keizer, T.H. and Emvalomatis, G. (2014) Differences in TFP growth among groups of dairy farms in the Netherlands. *NJAS – Wageningen Journal of Life Sciences,* Volume 70–71, pp. 33–38.

Kempen, M. *et al.* (2011) Spatial allocation of farming systems and farming indicators in Europe. *Agriculture, Ecosystems and Environment,* Volume 142, pp. 51–62.

Khayri, S.,Yaghoubi, J. andYazdanpanah, M. (2011) Investigating barriers to enhance entrepreneurship in agricultural higher education from the perspective of graduate students. *Procedia Social and Behavioral Sciences,* Volume 15, pp. 2818–2822.

Kirwan, J. and Maye, D. (2013) Food security framings within the UK and the integration of local food systems. *Journal of Rural Studies,* Volume 29, pp. 91–100.

Lal, R. (2013) Food security in a changing climate. *Ecohydrology & Hydrobiology,* Volume 13, pp. 8–21.

Lans, T. *et al.* (2014) Searching for entrepreneurs among small business ownermanagers in agriculture. *NJAS – Wageningen Journal of Life Sciences,* Volume 68, pp. 41–51.

Leck, C., Evans, N. and Upton, D. (2014) Agriculture – Who cares? An investigation of 'care farming' in the UK. *Journal of Rural Studies,* Volume 34, pp. 313–325.

Lee, H. (2012) How food secure can british cities become?. In: *Sustainable Food Planning: evolving theory and practice.* Amsterdam: Wageningen Academic.

Lee, H. (2015) *Urban entrepreneurship as a contributor to food security.* Leeuwarden, Wageningen, in press.Lee, H. (2016) Educating for food security in the UK: Planning for an uncertain future. *Local Economy,* in press.

Lockett, N., Jack, S. and Larty, J. (2012) Motivations and challenges of network formation: Entrepreneur and intermediary perspectives. *International Small Business Journal,* Volume 31(8), p. 866–889.

Mawois, M., Aubry, C. and Le Bail, M. (2011) Can farmers extend their cultivation areas in urban agriculture? A contribution from agronomic analysis of market gardening systems around Mahajanga (Madagascar). *Land Use Policy,* Volume 28, pp. 434–445.

Mrva, M. and Stachová, P. (2014) Regional development and support of SMEs – how university project can help. *Procedia – Social and Behavioral Sciences,* Volume 110, pp. 617–626.

Mumby-Croft, R. and Brown, R.B. (2006) SMEs, Growth and Entrepreneurship: The Steady Rise and Precipitous Fall of Seaking. *The Journal of Entrepreneurship,* Volume 15, pp. 205–217.

Nieuwenhuis, L.F. (2002) Innovation and learning in agriculture. *Journal of European Industrial Training,* Volume 26(6), pp. 283–291.

Norman, D. (2015) Transitioning from paternalism to empowerment of farmers in low income countries: Farming components to systems. *Journal of Integrative Agriculture,* Volume 14(8), pp. 1490–1499.

Nost, E. (2014) Scaling-up local foods: Commodity practice in community supported agriculture (CSA). *Journal of Rural Studies,* Volume 34, pp. 152–160.

Organisation Economic Co-operation & Development & European Commission (2012) *Policy Brief on Youth Entrepreneurship. Entrepreneurial Activities in Europe.,* Luxembourg: Publications Office of the European Union.

Petersen, B. and Snapp, S. (2015) What is sustainable intensification? Views from experts.. *Land Use Policy,* Volume 46, pp. 1–10.

Pihie, Z.A. and Bagheri, A. (2011) Teachers' and Students' Entrepreneurial Self-efficacy: Implication for Effective Teaching Practices. *Procedia – Social and Behavioral Sciences,* Volume 29, pp. 1071–1080.

Poole, D. (2001) Strategically managing entrepreneurialism: the Australian University experience.. *Higher Education Quarterly,* Volume 55(3), pp. 306–340.

Rodriguez, D. and Sadras, V.O. (2011) Opportunities from integrative approaches in farming systems design. *Field Crops Research,* Volume 124, pp. 137–141.

Roege, P.E. *et al.* (2014) Metrics for energy resilience. *Energy Policy,* Volume 72, pp. 249–256.

Rosen, G. and Razin, E. (2007) The College chase: Higher Education and Urban Entrepreneurialism in Israel.. *Tijdschrift voor Economische en Sociale Geografie,* Volume 98(1), pp. 86–101.

Runciman, B. (2012) Nature versus nurture in entrepreneurialism. *ITNOW,* September, pp. 18–20.

Salar, M. and Salar, O. (2014) Determining pros and cons of franchising by using swot analysis. *Procedia: Social and Behavioral Sciences,* Volume 122, pp. 515–519.

Sondari, M.C. (2014) Is Entrepreneurship Education Really Needed?: Examining the Antecedent of Entrepreneurial Career Intention. *Procedia – Social and Behavioral Sciences,* Volume 115, pp. 44–53.

Sorrell, S. *et al.* (2012) Shaping the global oil peak: A review of the evidence on field sizes, reserve growth, decline rates and depletion rates. *Energy,* Volume 37, pp. 709–724.

Spiller, K. (2012) It tastes better because …. consumer understandings of UK farmers' market food. *Appetite,* Volume 59, pp. 100–107.

Starr, A. (2010) Local Food: A Social Movement? *Cultural Studies – Critical Methodologies,* Volume 10(6), p. 479–490.

Stenholm, P. and Hytti, U. (2014) In search of legitimacy under institutional pressures: A case study of producer and entrepreneur farmer identities. *Journal of Rural Studies,* Volume 35, pp. 133–142.

Strohbach, M.W., Kohler, M.L., Dauber, J. and Klimek, S. (2015) High Nature Value farming: From indication to conservation. *Ecological Indicators,* Volume 57, pp. 557–563.

Taleb, N.N. (2013) *Antifragile: Things that Gain from Disorder.* London: Penguin.

Tittonell, P. and Giller, K.E. (2013) When yield gaps are poverty traps: The paradigm of ecological intensification in African smallholder agriculture, *Field Crops Reseearch,* 143, pp. 76–90.

Woods, P.A. and Woods, G.J. (2009) Testing a typology of entrepreneurialism. Emerging findings from an Academy with an enterprise specialism. *British Educational Leadership, Management & Administration Society (BELMAS),* Volume 23(3), pp. 125–129.

Yokoyama, K. (2006) Entrepreneurialism in Japanese and UK universities: Governance, management, leadership, and funding. *Higher Education,* Volume 52, pp. 523–555.

Young, J.E., Griffin, M.J., Alford, D.V. and Ogilvy, S.E. (2001) *The TALISMAN & SCARAB projects,* London: Defra.

16 Experiences of ten European cities collaborating toward sustainable food governance in an URBACT network

François Jégou and Joy Carey

Introduction

The project and its partners

The URBACT II thematic network Sustainable Food in Urban Communities – Developing low-carbon and resource-efficient urban food systems (URBACT, 2012–2015) brings together ten European cities looking for joint, effective and sustainable solutions to develop low-carbon and resource-efficient urban food systems (Figure 16.1). The URBACT process involved regular transnational exchanges between the ten cities over a period of three years and provided a framework for each city to establish a Local Support Group of key stakeholders in order to collaboratively build sustainable food governance and a related local action plan. The cities of our network offer a fascinating variety in terms of (1) demographics and scale of urbanisation; (2) land and territory; (3) food culture; and (4) levels of engagement in food system sustainability.

Finding angles to view the urban food system and sustainability

Three themes of 'Growing', 'Delivering' and 'Enjoying' were selected by project partners to approach the complexity of the food system more simply, and to find a way to organise the many vibrant and heterogeneous experiences of the ten cities. The three themes were used to collect and review practical case studies of existing work in each city. The theme of 'Growing' explores all possible ways to grow food near or in the city. The theme of 'Delivering' explores ways to distribute, share and procure food within the city. The theme of 'Enjoying' explores how people in the city can embrace a sustainable, happy, healthy and vibrant food culture in canteens and households. In addition, project partners addressed three cross-cutting issues: 'Governance, synergies & local system'; 'Social Inclusion, jobs and economics'; 'Carbon emissions and resource efficiency'.

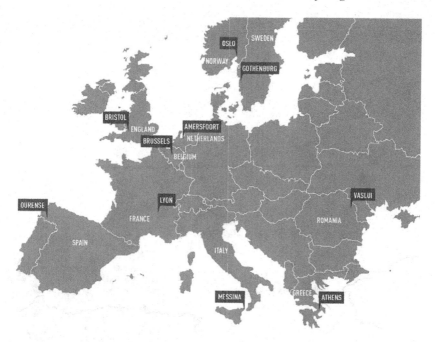

Figure 16.1 The URBACT thematic network Sustainable Food in Urban Communities

Note: The three-year exchange and collaboration project involves 10 European cities including
  Brussels Environment of the Brussels Capital Region (Lead Partner), Bristol City Council, City
  of Messina, the Municipality of Amersfoort, the City of Lyon, the City of Gothenburg, Vaslui
  Municipality, Ourense City Council, City of Oslo and Athens Development and Destination
  Management Agency SA.

Finding an action-focused framework for sustainable food in cities

Initiatives that address the need for sustainable food in cities can contribute to
supporting a lively and diverse local economy (jobs and skills), to creating a better
environment (green spaces, urban design, reduced greenhouse gas emissions), and
to supporting more health and wellbeing amongst the population (inspiring behaviour change, making it easier for people to make better choices). In order to
achieve these kinds of benefits, food has to be put firmly on the city governance
agenda. For that reason, over the course of the project the focus on 'Growing',
'Delivering' and 'Enjoying' has gradually shifted towards more of an action-
oriented and organisation-based focus of 'Sustainable entrepreneurship', 'Citizen's
resilience' and 'Food governance' (Figure 16.2). Creating space for sustainable food
systems is a practical and physical challenge in terms of finding available land
within the city and its outskirts, and is also about creating space for food in the
broader economic, legal, cultural and lifestyle context:

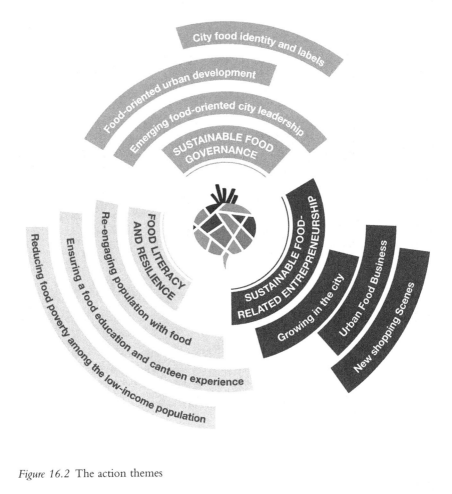

Figure 16.2 The action themes

- Creating space for emerging economic activities and food-related sustainable entrepreneurship in the city;
- Creating space for good and healthy food in everyday life of more resilient citizens;
- Creating space for food in the governance of the municipality with traditionally limited authority in this area.

These action themes are discussed in more detail in the following sections with illustrative examples from the ten partner cities.

What can cities do?

Getting started

The question is how can a city positively influence its food system? A good first step is to develop better understanding and establish dialogue. This can then feed into a longer-term 'sustainable food planning process' which may take several years or indeed be adopted as part of an on-going food strategy and policy agenda. There are many different ways of getting started. Bristol commissioned a baseline audit study called 'Who Feeds Bristol?'. Stakeholder meetings, organised public events and discussions and participatory scenario exercises all help to stimulate debate and establish a new 'sustainable food conversation'. The most important point is that it has to be both a process and a partnership approach.

Sustainable food-related entrepreneurship

For the participating cities, creating space for food relates firstly to enabling emerging economic activities and food-related sustainable entrepreneurship. This includes: securing urban land/space to enable growing food in the city; fostering the emergence of new urban food businesses with innovative income generation models; and facilitating the development of a new shopping scene with innovative models of intermediation between local food producers and city dwellers.

Growing food in the city

The trend of reclaiming urban space to grow food, observed in participating cities, often involves initiatives born from very different motivations. Urban gardening might start as a production enterprise or a recreational activity to engage people with food. Several cities encourage the development of peri-urban farming to reinforce local supply. The 'Feed Bristol' project combines both growing food and caring for wildlife (Box 16.1). It may also start as an arts project, an educational initiative, a pretext for new social connections or an activity facilitating social inclusion of marginalised groups, as for instance the Prés Senty project in Lyon (Box 16.1). These different motivations tend to blur and most growing initiatives are multipurpose. This multipurpose characteristic is fundamental to understanding the value of food gardening in cities, which is clearly limited in terms of scale and volumes of food produced. Challenges include re-engaging disadvantaged population groups in cooking fresh food from scratch instead of living on more expensive industrialised ready-made food, or shifting from overconsumption of junk food to more healthy choices. Reconnecting urban citizens with food growing in a natural environment seems to provide solutions. In particular, the experience of growing food for oneself can be life-changing. Therefore, participating cities acknowledge the indirect social and economic benefits of recreational, educational or even symbolic urban food gardening as much, if not more than the actual quantities of food produced by urban agriculture initiatives. Within a city the benefits of growing food should be assessed taking into consideration the whole urban system and accounting for indirect and collateral positive effects as well as direct contributions in terms of effective agriculture production.

Box 16.1 Projects from Bristol and Lyon

Feed Bristol project and Sims Hill shared harvest, Bristol

Situated in the outskirts of north Bristol on a 7 acre piece of land, the educational wildlife-friendly food growing project 'Feed Bristol' is twinned with 'Sims Hill Shared Harvest'. Feed Bristol provides opportunities for volunteers to grow food and care for wildlife. Sims Hill is a community supported agriculture scheme with 65 members divided in 3 categories: growing members, vegetable sharing members and supporting members. The second category is particularly interesting: engaging people with nature and food is not always easy when they do not have time to take care of their own individual allotment. Vegetable sharing members help for four hours a week during six months and in exchange get access to vegetables year round.

Pré Santy inclusion garden, Lyon

The 'Pré Santy' is a vegetable garden aimed mainly at improving social inclusion in a difficult social housing area in the south-east of Lyon. The gardening activities are more a pretext than an aim, but it is an interesting example of promoting sustainable food amongst an underprivileged population. The garden occupies a small piece of land surrounding a parking lot but it is large enough to enable 20 families to experience eating the vegetables they produce from time to time and to organize more than 20 neighbourhood events around self-grown food per year.

Urban food businesses

New entrepreneurship related to sustainable food in partner cities and in nearby suburban areas gives rise to new and innovative income generation or value creation models based on hybrid partnerships. It often involves a creative mix of social innovation, public-private partnerships, shared and collaborative economy beyond the classical social contract between public authorities, civil society, businesses and citizens. The city of Ourense for instance in the Spanish region of Galicia promotes an interesting project built around short food circuits (Box 16.2) fostering both supply and demand at the same time: the municipality facilitates the emergence of youth entrepreneurship in market gardening in nine villages around the city generating jobs and suburban local food supply. In parallel, the municipality encourages customer demand through the refurbishment of the old traditional markets of the city as an attractive fresh food hub with cafés and small restaurants.

The Cuppari agriculture high school in Messina (Box 16.2) is a good example of an even more complex and interwoven multifunctional model. The school's integrated economic strategy promotes high-quality regional food and increased job opportunities. It connects the use of winery teaching equipment of the school to produce premium wine, the development of a side business of selling wine to compensate for public school budget shortages, the promotion of regional products creating a wine tasting area in the school to increase local tourism and the enrichment of students' experiences and skills.

Promoting these new multi-stakeholder income generation and value creation models requires cities to find space, not only physical space to establish the new

---

Box 16.2 Projects from Ourense and Messina

'Gardens Bank': recovering food production areas in the outskirts of Ourense

Since 2009, Ourense city council has been encouraging different projects pooling unused land for gardening. The organisation of such an innovative project, called 'Gardens Bank', relies on cooperation between landowners (in particular people who moved away from these areas or elderly people who cannot take care of their land and therefore generate risk of fire) with potential producers (in particular young people suffering from high levels of unemployment). The municipality proposed to relieve the landowners of the fire fines if they lend the land they do not use to young entrepreneurs for market gardening. In order to ensure market access and boost this suburban food production the municipality focuses on the rehabilitation of two traditional markets as 'food hubs' for the population of the city, developing social life around the traditional markets, enhancing local food use among the small bars and restaurants around these markets, and offering free public transport access for sellers and entrepreneurs coming to offer their horticulture products at the local markets. Likewise, the municipality attempts to promote self-consumption in sectors of the population with low incomes, especially among producers themselves. Today, Ourense has 12 orchards in production and a waiting list of 24 citizens!

Faro Doc from the Cuppari High School, Messina

Passionate staff from the Agricultural Institute 'Cuppari' devised a strategy to address the school budget shortage: from teaching wine production they have now extended their activities to produce quality wine and have recently launched a new faro DOC (of controlled origin) 'S. Placido' high-quality wine with the first 12,000 bottles sold in 2014. They have also opened a local wine tasting area and plan to offer direct access for tourists from cruise ships passing through the Messina strait.

production businesses but also a widened understanding of the public authorities' role and administrative culture. The municipality operates as a broker in the city to facilitate partnerships, to experiment with them and to secure public equity, fair exchanges and forms of co-responsibility between involved players (Jégou & Bonneau, 2015).

New shopping scenes

The hybrid and multifunctional nature of new sustainable food urban entrepreneurship induces a renewal of the shopping scene in the partner cities. This concretely leads to a disintermediation (cutting out the middle man) between producers and users and to a growing level of customer involvement in the quality control and delivery of the services. This new shopping scene ranges from well-known and developing forms of bulk food purchasing groups (GASAP in Brussels, AMAP in Lyon, GAS in Messina, etc.) to a change in the traditional model of supermarkets, adapted by new entrepreneurs to welcome small local producers and ensure affordable quality food (i.e. Plus Supermarket in Amersfoort). Half way between these extremes, suburban producers also create their own distribution models to nearby urban areas. The Landwinkels' network of shops at the farm gate in the Netherlands (Box 16.3) has developed a local cooperative brand supporting farmers in setting up professional shops to sell their products directly on site. An example of a new hybrid form of retail is La Superhalle in Lyon, combining several related businesses in a former storage building of on 800 m2 within an industrial neighbourhood in the outskirts of the city: an organic grocery area furnished with shelves and bulk dispensers, a farmers' market with different producers serving on a weekly rota arrangement, an organic restaurant and take away and a vegetable box scheme delivery point (Box 16.3). In some partner cities an increasing number of vegetable basket or online order schemes compete in flexibility and attractiveness for consumer attention, providing information about the local producers to create a human connection. This emerging shopping scene is changing food distribution opportunities in the partner cities. Short circuits, direct purchase by the producers, sales of organic or traditional farming products tend to increase the average food quality and reduce packaging waste. However, this kind of localised delivery in the city tends to be less optimised. The customer logistics (the duration and routes followed by individuals to visit different shops, collect their purchase and bring them home) can be more complex and thus potentially reduce the environmental benefits.

Reflections on sustainable food-related entrepreneurship

Of particular interest is the innovation and multifunctionality of these examples. They bring additional economic, environmental and social benefits and services that contribute to the lives of citizens and to the city's distinctiveness. They bring increased diversity of retail models; increased connection between food producers and customers; and new local markets for food producers. The importance of the

---

Box 16.3 Projects from Amersfoort and Lyon

Landwinkels, shops at the farm gate, Amersfoort

Landwinkel is a national organisation that supports farmers in opening shops at the farm gate. Landwinkel works as a cooperative, pooling products from local farms so that each outlet offers not only its own products produced onsite but all the products produced locally by the network of local farms. Landwinkel also provides commercial and merchandising support (i.e. branding, packaging, shop design, information and advertisement material), greatly facilitating farmers' ability to have their own shop and attract customers to the outskirts of the city (where they are mainly located).

Super Halle, Lyon

The Super Halle is an innovative concept of a grocery shop based on sustainable food. It combines in one place: an organic shop, a producer shop (fresh produce direct their farms), a restaurant and a food supply/distribution hub. Those businesses, run by four independent companies or cooperatives, are supporting each other. The restaurant and organic shop are supplied by the producers, who also do shifts at the shared cash desk. The organic shop doesn't sell what is available at the producer shop, instead offering complementary foodstuffs. This multifunctional system, based on innovative cooperation and coordination schemes, provides an interesting new business model and a unique shopping experience for their customers.

---

entrepreneur cannot be overemphasised. Policy and voluntary activism alone are insufficient to create food system reform.

Food literacy and resilience

'Food literacy' is about ability and understanding in relation to food. Organisations promoting this concept define it in terms of personal behaviour: the ability to organise one's everyday nutrition in a self-determined, responsible and enjoyable way and as understanding the impact of food choices on health, the environment and community.

'Resilience' is not such a well-defined term. In this context we have used it to describe the capabilities of citizens to maintain a balanced and quality food diet in challenging situations such as economic crisis and low income or perturbations in the usual food delivery system.

Creating space for food in the city therefore also relates to cultural and daily behaviour and practices:

- Convenience food has become asocial standard or new social norm.
- Mainstreaming healthy and quality food calls for re-engaging the population with food, in particular by dedicating some time and securing enough skills to prepare and cook their own food.
- It calls also for ensuring a food education and canteen experience.
- Spaces should be created at school and elsewhere since, in an increasing number of situations, it is less likely to take place at home.
- Finally it requires solutions within an urban food system to reduce food poverty among low-income population groups.

Re-engaging the population with food

The participating cities represent different dimensions of the European food culture landscape. They range from weaker food cultures invaded by junk food and agro-industrial products to richer food cultures, but which are often blind to the progressive loss of skills and interest in food by the younger generations. In all these situations, a more or less growing part of the population is showing a strong disengagement with food. Sophisticated processes are required in order to help people to re-engage in cooking and in appreciating healthy and quality food.

The Brussels Capital Region distributed thousands of free grow-your-own kits to anyone interested in setting up an environmentally friendly kitchen garden in open soil, on a balcony, terrace or in pots. The Square Food School in Bristol (Box 16.4) is a particularly illuminating example. A skilled and motivated chef opened a food café next to the Square Food School's training kitchen, offering a premium quality menu compatible with the training courses operating at affordable prices. All the required conditions seemed to be met to re-engage the local population with good food and with the pleasure of preparing it. But it did not work. Customers were decreasing. The food café reintroduced a share of junk food and soft drinks in its menu. It is now offering an improbable mix of highly healthy and highly unhealthy items: a form of 'transition food' to retain interest of local customers while slowly influencing and improving their food habits. The Food Challenge project (Box 16.4) has been piloted by the commune of Etterbeek in Brussels with low-income households who were willing to find out more about food sustainability but were unsure how this could fit into their daily lives and into their budget. The Food Challenge is a transformative process involving a sample of families in a six-month series of participative food training sessions. The project was assessed as very effective but the costs incurred are judged to be very high given that it only prompted changes in the way of eating of 12 families, who are now willing to take on the role of sustainable food ambassadors at events in the municipality. A photo exhibition, a website and a brochure with recipes and advice given by the participating families helped reach out to a wider public. Still, the Food Challenge is also a challenge in terms of scaling-up such initiatives in a cost-effective way to engage more families. In both cases and in many other serious and in-depth initiatives aiming at re-engaging people with food or bringing in environmental considerations in food choices, lifestyles changes are effective and rather long lasting. However, these initiatives also require a high level of involvement from their promoters. The complex transformative processes required to change habits and catch-up with a lost – or disappearing – food culture, are expensive. The costs incurred for municipalities to replicate these processes at city scale are inevitably perceived as barely viable.

Box 16.4 Projects from Bristol and Brussels

The Square Food Foundation cookery school and kitchen, Bristol

The Park is home to the Square Food Foundation and its training kitchen. Knowle West is one of the more vulnerable neighbourhoods in Bristol where poor eating habits are quite entrenched. An attempt to establish a good quality wholesome food café next to the training kitchen failed through lack of interest from customers. The offering is now 'transition food': a mixture of wholesome and 'junk food' so it stays to some extent familiar and therefore accessible to local people. This, together with open cooking classes for customers, provides a progressive pathway to sustainable and healthy food.

Food Challenge for Families, Brussels

To support grassroot projects about sustainable food, almost every year Brussels Environment launches a call for proposals about sustainable food. Several calls for proposals have been focused on low-income households in order to increase sustainable food availability.

One of the supported projects was, 'Food challenge' of Etterbeek (one of the 19 municipalities of the region). The idea was to coach a small number of households to create a positive momentum around sustainable food by empowering families to become ambassadors on the subject. Low-income households not yet aware about sustainable food issues were given priority.

Over a period of 6 months, 12 households were involved every fortnight in activities such as cooking classes and tasting workshops. Behaviours and views on food quickly evolved. A photo exhibition, a website and a brochure with recipes and advice given by the participating families helped reach out to the wider public. This interesting project was much appreciated by participants and has been repeated a second time. The high cost of the intense coaching activities, however, limits the number of families that can be accompanied.

Ensuring a food education and a positive canteen experience

Kids and youngsters have fewer and fewer opportunities to learn about food and cooking at home due to a combined lack of knowledge, time and interest from their parents. The food culture and related skills of growing, preparing and tasting food are disappearing from household environments. Partner cities of the network all report numerous initiatives at school level to address this gap and re-integrate food education through two main entry points: first, the improvement of the quality and sustainability of the food served in the canteens – this strategy is leveraging on public procurement to encourage the sustainable transition in all parts of the food chain. It is also ensuring at least one quality balanced meal offered to each child per day. Second, the inclusion of food as a proper topic in teaching curriculums – this strategy includes the implementation of educational gardens and cooking activities in the schools and in some cases visits to farms. If these two entry points are combined (as for instance in the UK Food for Life Partnership approach for schools) it maximises the benefits gained by the children. Furthermore, while the start-up is demanding, once the schools are fully engaged, the process is self-perpetuating as this is a whole school integrated approach, involving all staff.

The case of Lundby school in Gothenburg (Box 16.5) shows the importance of the involvement of canteen staff and chefs: they catalyse change, motivate pupils and challenge in a creative way the rules and legislation around canteens which can appear restrictive. Specific food education processes are also designed for kids as for instance the Geitmyra Culinary Centre in Oslo (Box 16.5). It offers weeks of immersion for kids to perform the complete cycle of growing, transforming, cooking and eating their food.

Beyond growing interest in food issues and in cooking practices, the challenge of food education at school is to raise the importance of food in daily life, as a pleasure and as an element of quality of life. In order to ensure quality food becomes an important issue (and stays important) for citizens it is key to raise the tasting capabilities of kids. Their ability to recognise good-0quality food by taste (rather than only from the label) will ensure that they will appreciate quality food and continue to request it throughout their lives as adult consumers.

Reducing food poverty among the low-income population

The first concern of partner cities, when facing food poverty issues, is to try to redistribute food that would otherwise be wasted. This is happening in the retail sector and also in the kitchen. Organisations like FareShare in the UK and many local non-profit organisations recover and redistribute unsold food, helping to rebalance a malfunctioning food system in situations of low or no income.

Beyond reducing food waste, partner cities stressed citizens' food resilience in urban contexts can be improved by securing access to land. The municipality of Athens is involved in reclaiming unused land available within the city and turning it into urban farms (Box 16.6). This structured solidarity project aims to both create jobs for unemployed people with low qualifications in the city and generate an urban production of food for poor households hit by the crisis.

In Lyon, the social and solidarity grocery shop La Passerelle d'Eau de Rebec (Box 16.6) shows another mechanism to increase citizens' resilience in relation to food access. The grocery is a non-profit business. It collects benefits from 'solidar-

---

Box 16.5 Projects from Gothenburg and Oslo

Lundby local administration is serving almost 50 per cent organic meals in public canteens, Gothenburg

The local administration in Lundby (part of the city of Gothenburg) is leading the way in serving organic food. They are now serving almost 50 per cent organic meals in canteens for schools and for homes for elderly. A couple of months ahead of schedule, the city of Gothenburg qualified in the league of organic food in Sweden (Ekomatsligan) because they are on average serving 33 per cent organic food in the local administrations in Gothenburg (canteens throughout the city). This shows that the local administration of Lundby is at the leading edge. Organic food is a priority and we work with seasonal food, it takes more hours to cook food from scratch. We want to aim even higher, says Johan Fogler, divison manager and chef in the Elias kitchen.

Geitmyra culinary centre for children, Oslo

Geitmyra culinary centre for children is a non-profit foundation established by the food writer and TV-cook Andreas Viestad. Based in the buildings of a former farm near Geitmyra allotment garden in Oslo, the place has been refurbished to host school children every day of the week, in addition to kindergartens visits and classes for adults who work with children and food. It also offers evening courses and open farm arrangements. It hosts 30–35 classes per year for an entire week each. Kids have an immersive experience on a farm where they can experience the origins of food by feeding and slaughtering chickens, growing and taking care of the organic vegetable garden, baking bread, smoking and conserving fish, making yogurt etc. and eating the products of their labour.

---

ity' customers purchasing quality food and it uses these benefits to enable 'beneficiary' customers of the same neighbourhood to also buy quality food, but at a lower price. In both cases, the municipalities involved are trying to increase citizens' food resilience. They are adapting traditional models of resilience (i.e. based on complementary self-production and mutual help within a community) to current urban environments.

Reflections on food literacy and citizens' resilience

Two aspects are highly pertinent and represent a serious challenge to cities: first, how to successfully enable citizens to have new learning experiences that influence their understanding, confidence and ultimately their behaviour about food choices; and second, how to provide the practical infrastructure to improve physical access

> **Box 16.6** Projects from Athens and Lyon
>
> ## Eleonas urban farm project, Athens
>
> The Municipality of Athens is supporting a large action towards reclaiming available urban space for growing food. It is implementing real urban farms on the larger pieces of land and urban gardening on smaller plots. Eleonas is a neighbourhood of Athens with mixed industrial and commercial areas. A large brownfield is planned to be turned into an urban farm (municipal property of approximately 20,000 m2). The project is about to start and foresees benefits both in terms of creation of jobs for unemployed people and of urban production of food for poor households. At the same time, small-scale urban farming is being advanced through the identification of plots of land that can be used for cultivation aiming at educational and symbolic functions (e.g. local edible wild vegetation, vegetable, herb and tree gardens).
>
> ## Social and solidarity grocery shop, Lyon
>
> The original grocery La passerelle d'Eau de Rebec is a social business enhancing social inclusion with a dual pricing system. The grocery works with two types of clients: 'beneficiaries' with low incomes accessing mainly food charity stock but also fresh organic food sold to them below the market price, and 'solidarity clients' attracted by organic food sold at fair but full market price to them. Both evolve in the same space: the solidarity clients are providing better margins to the shop. These margins are reinvested in order to give beneficiary clients a discount on organic, local and fresh food.

to good healthy food in affordable ways. Rarely is one sufficient without the other. New habits need to be developed through educational activities. However, in situations of low income, education alone is simply not enough.

Food governance and the city agenda

Creating space for food in the city relates ultimately to the governance itself. Municipalities have traditionally had limited authority in the area of food. Cities involved in the URBACT network begin to develop a specific governance striving towards sustainable and quality food. They seek to raise awareness amongst decision makers on the strategic dimensions of food in an urban context and to foster proper food-oriented city leadership. Partner cities also consider food an asset. They explore possible synergies between food and other sectors of the city, overcoming administrative silos. A food-oriented urban development is emerging. Finally, the partner cities focus on food culture at the territorial level. They endeavour to pool together local food assets in a coherent way and create or recreate a city food identity and label.

Emerging food-oriented city leadership

Cities usually do not have any (or limited) official authority in terms of food. Food issues are generally dealt with at higher regional and especially national and international levels. But food brings together a number of local insidious problems that present challenges to municipalities: food poverty, food-related diseases, unhealthy diets, unequal access to quality food, city food supply vulnerability, etc. These many problems with their complex interdependencies and rebound effects should urge motivate cities to engage in food policies.

The City of Bristol, inspired by Canadian and American examples, created a Food Policy Council (Box 16.7). This council is an informal advisory group of a dozen key stakeholders. It investigates food issues at city level and it is available to advise local policy makers on food-related issues. It summarises its function and role as 'validating, influencing, connecting, communicating and creating visibility' of all aspects of the sustainable food agenda.

Partner cities also aim to tackle food as an issue in all branches of the municipality. The Brussels Capital Region for instance explored a joint policy making process driven by the ministers of Economy and Environment in collaboration with local stakeholders: the Job-Employment Alliance (Box 16.7). The overarching aim is to generate food-related entrepreneurship and employment through the development and greening of the sector by jointly identifying barriers and targeted actions needed to enable such development.

Partner cities of the network show a growing involvement in addressing sustainability in the food system. Municipalities integrate the sustainable food topic in various levels of their administration. They make use of their purchasing power at a strategic level to stimulate sustainable transition of their food providers. Their very involvement in the Sustainable Food in Urban Communities URBACT thematic network shows their strong interest in exchanging policy practices, in building a Local Action Plan and in creating city leadership on the topic.

Food-oriented urban development

Urban planning in modern cities hardly ever takes food issues into consideration. Partner cities of the network reveal their lack of background and preparation on the topic. In particular, they generally lack information and studies on food in the urban perimeter. For example, who are the stakeholders? What are the flows of food supply from the urban perimeter entering the city? How is this food distributed to reach the consumers' plates? How are leftovers disposed of? What are the problems and where can unmet opportunities be found?

The *Who feeds Bristol?* report (Box 16.8) covers all these issues. It suggests a range of strategic orientations and priority actions to operate a transition towards a more healthy and sustainable food system. This report was acknowledged as an inspiring model by the partner cities. Most of them have undertaken similar *Who feeds the city?* studies covering their own territory.

In parallel, partner cities involved in the URBACT process set up a 'Local

Box 16.7 Projects from Bristol and Brussels

Bristol Food Policy Council

The Bristol Food Policy Council was launched in March 2011, as the first Food Policy Council in UK. It brings together a dozen stakeholders from diverse food-related sectors to examine how the food system is operating locally and to develop recommendations on how to improve it. Bristol Food Policy Council has been established as an independent body including representative from the local government and a board of local key players. Internationally Food Policy Councils are educating officials and the public, shaping public policy, improving coordination between existing programmes, and starting new programmes, mapping and publicising local food resources, creating new transit routes to connect underserved areas with full-service grocery stores, persuading government agencies to purchase from local farmers, and organising community gardens and farmers' markets, etc.

Job creation policies in urban agriculture and sustainable food businesses, Brussels

A study of the economic potential of businesses related to sustainable food estimated that about 2000 jobs could be created by 2020, particularly in the field of food production. Regional ministers for Environment and Economy and their administrations are working together to develop this sector in collaboration with stakeholders through various mechanisms: the 'Jobs-Environment Alliance' (www.aee-rbc.be); regional European Regional Development Fund resources giving priority to sustainable food projects; and an upcoming transversal action plan on sustainable food.

Support Group' of stakeholders from the beginning of this programme, building either on groups or networks that already existed or starting from scratch. The Local Support Group aims to develop a proper sustainable food strategy. It articulates a set of priority actions already in progress in the city with actions inspired and adapted from other partner cities of the network. These actions should be organised and presented into a Local Action Plan, which is approved by the municipality by the end of the URBACT process in mid-2015. The Local Action Plan should embody the emerging sustainable food policy of the city.

Among the rich set of actions generated in the network, the programme of rebuilding food markets by the city of Vaslui (Box 16.8) is a characteristic example. A large foreign international supermarket brand is currently settling in the city. At the same time the municipality is building brand new food markets offering similar facilities (i.e. large parking underground or on the roof top; similar size hosting many market stalls and shops etc.) and combining them with traditional market assets (hosting quality fresh food with specific clearly identified areas reserved for local farmers and controlled by a laboratory established within the market). The partner cities illustrate how the URBACT process can be used to start or enrich food-oriented urban development. The process helps to bring together and strengthen existing opportunities for a sustainable food transition, and to explore and activate potential synergies working towards a more robust local sustainable food system.

Box 16.8 Projects from Bristol and Vaslui

Who feeds Bristol? report

A large and extensive report was commissioned both by Bristol City Council and Bristol National Health Service. The report was led an independent expert, in collaboration with the food community in Bristol. It draws an in-depth and exhaustive overview of food issues in Bristol as a basis for a food systems approach. The report identified a range of challenges and actions that have provided direction in Bristol, now articulated by the Good Food message, the framework for food planning in Bristol.

Food market programme in Vaslui/Central market, Vaslui

A new retail market was built from scratch on a former derelict market site, with local budget funds (approx. EU€3.5 million) in the centre of the city as an energy efficient building. Work started in 2012 and finished in September 2014, and the market is administered by the local authority. It is endowed with the high-level technical means to facilitate the direct sales of local products coming from the small-sized land holdings of Vaslui.

The market is divided into five well-designed areas: quality control laboratories for food safety and security (which plays an important role in increasing the trust of people), fruit and vegetables, fish products, meat and diary products. Local producers are the main target of the market and they have special designated areas with special rent conditions in order to encourage local production of best-quality products, shorten the food chain and reduce CO_2 emissions.

City food identity and labels

Ways to make food issues more visible and influence behavioural change at city level include local training and events, the use of awards, recognition schemes or labels, and the collective identification of positive examples. These can for instance inspire and encourage the purchase of fresh seasonal and local food; reduce food waste; and encourage circular economy thinking. Labelling and recognition schemes need to be simple and aimed also at 'non-green' audiences and help people to make a small shift. Cities can use such schemes to good effect.

For instance, the city of Lyon, well known for its high-level cuisine and vibrant cooking culture, intends to raise awareness on food sustainability. Its strategy is to capitalise on its already-successful label of 'Lyon Fair and Sustainable City' (Box 16.9) to stimulate the sustainable food sector.

For national cultures where food is less of a focus, an event, a campaign or a process that raises the profile of food issues and stimulates activity can provide a basis on which to build. The city of Amersfoort deploys large resources to stimulate food culture through various events to raise more interest in food among the emerging food activist communities and beyond. In 2012 the city was awarded 'Capital of Taste' of the Netherlands (Box 16.9). The municipality built on this opportunity and subsequently on the participation in the URBACT network to keep the momentum and continue the process of raising more interest in food locally.

Collective identities, labels or awards can be helpful positive tools, at least as a capitalisation process in order to give coherence and visibility to the issue of sustainable food. Exposure to these kinds of initiatives or schemes can help to create aspiration and stimulate the emergence of more initiatives that relate to sustainable food within the network.

Box 16.9 Projects from Lyon and Amersfoort

Lyon Fair and Sustainable City label

The Lyon Fair and Sustainable City label was set up in 2010 in order to create a community of sustainable practices among the trade people in town. The label is characterised by transparency of selection criteria and progressive improvement processes for those who are not yet eligible. Due to the city's culinary cultural background (40 per cent of the businesses belong to hotels, restaurants and cafés sector), the food topic is very well represented among awarded entities. The label stimulates cooperation between them and between food-related and non-food related fair and sustainable activities in the city. The aim is to reach 200 active members and strengthen the activities of the network.

Amersfoort Capital of Taste 2012

Throughout 2012, the designation of the region of Amersfoort as 'Capital of Taste' was the occasion to organise a large variety of recreational, educational and popular activities around food. Together with several regional municipalities and citizens involved in local food actions, Capital of Taste was a good way to promote regional food and address the connection between health, seasonal, fair and pure taste. Every year the elected Capital of Taste focuses on a signature product. In the case of Amersfoort and the surrounding municipalities, the typical Dutch potato in all its refined and gastronomic aspects was the focus product. Capital of Taste 2012 culminated with the national Week of the Taste in early October, with almost 700 activities in the Netherlands.

Reflections on food governance and the city agenda

The challenge that every city faces is first how to get the subject of sustainable food onto the city agenda and second how to keep it there. Solutions encompass a range of interventions research, infrastructure and skills investment, cultural activities etc. that together begin to make up what we are calling 'food governance'. These interventions help to embed new values and principles within city culture and the policies and strategies of the municipality, other stakeholder organisations and businesses. At the core is a new partnership approach to food system reform, between the municipality and other stakeholders. This process of working in partnership is not easy but is possibly the only way to make the informed and impactful stepchange at the scale and specificity that are required for a sustainable and low-carbon food system future.

Conclusion

Food emerges as a new burning concern for which the cities in the network – though committed – were not entirely prepared. The experience of the network shows that striving towards a local sustainable food system goes beyond the question of land use and availability within the city and its outskirts. A wider perspective is required addressing economic, legal, cultural and lifestyle questions. Finding and making space for food is therefore a very pertinent issue for cities – not just in terms of physical space for food entrepreneurship and self-production, but in terms of a widened understanding of the public authority's role and its administrative culture.

Cities are thus having to fight for space in terms of governance in order to facilitate economic activities including food production, but also new models of value creation or income generation that combine multiple functions in a hybrid mix of production, delivery and transformation (processing); change cultural representations of food, invest in infrastructure that helps to make good food more affordable and encourage a shift in daily practices towards more healthy and quality food choices; and adapt legal and administrative procedures and practices to enable partnerships and new forms of city food leadership at the scale of the urban territory, and also connected to food and agriculture policy at regional and national levels.

Partner cities are in the process of changing their governance to match the crosscutting and multidimensional nature of the food system. It is both a challenge and an opportunity for them to develop a more participatory and multilevel form of food governance.

Reference

Jégou, F. and Bonneau, M. (2015) *Social Innovation in Cities, URBACT II Capitalisation.* URBACT Publishing, Paris.

Index

Printed in the United States
by Baker & Taylor Publisher Services